Maple V Calculus Labs

Abi Fattahi

Whittier College

Brooks/Cole Publishing Company
Pacific Grove, CA

AF876507

Brooks/Cole Publishing Company
A Division of Wadsworth, Inc.

© 1992 by Wadsworth, Inc., Belmont, California 94002. All rights reserved. No part of this book may be reproduced, stored in a retrieval system, or transcribed, in any form or by any means—electronic, mechanical, photocopying, recording, or otherwise—without the prior written permission of the publisher, Brooks/Cole Publishing Company, Pacific Grove, California 93950, a division of Wadsworth, Inc.

Printed in the United States of America

10 9 8 7 6 5 4 3 2 1

Trademarks
Macintosh is a registered trademark of Apple Computer, Inc.; **IBM PC** is a registered trademark of International Business Machines Corporation; **Maple** is a registered trademark of Waterloo Maple Software. All other product names are trademarks or registered trademarks of their respective companies.

Sponsoring Editor: *Jeremy Hayhurst*
Editorial Assistant: *Nancy Champlin*
Production Coordinator: *Marlene Thom*
Manuscript Editor: *Andrea Olshevsky*
Cover Design: *Lisa Berman*
Typesetting: *Supplinc*
Printing and Binding: *Malloy Lithographing, Inc.*

Preface

In recent years, there has been much discussion about reform in the teaching of calculus, and computer-based instruction in mathematics in general. The idea is to use the capabilities of a so-called Computer Algebra System (CAS) to reduce the amount of time spent on routine calculations, thus providing more class time for discussing mathematical concepts and problem-solving techniques.

If used properly, a CAS can be of great assistance in explaining important concepts of calculus such as limits, derivatives, definite integrals, and approximations of functions by polynomials. Furthermore, they can be quite useful in pattern-recognition situations and discovery of formulas and relationships. Finally, for the more experienced user of mathematics, CAS are an extremely useful tool for solving problems involving mathematical applications. These lab sessions were designed with all three of those objectives in mind. The earlier sessions emphasize understanding the important concepts of calculus and discovery of patterns, whereas many of the later sessions demonstrate the use of a CAS in solving various problems involving lengthy calculations. In this manual we refer to **Maple V**, one of the most powerful CAS available today. We refer specifically to the versions of Maple that are designed for use on the Macintosh® or IBM-PC® personal computers and clones. Both the professional **Maple V** and the **Student Version of Maple V** (available from Brooks/Cole) are appropriate. Most of the concepts and procedures discussed are applicable to other versions of Maple, as well as other Computer Algebra Systems, however.

These notes are not intended for casual reading. Each lab session is designed to explain a concept or a process. Students are expected to read the explanations and instructions carefully while sitting in front of the computer, and whenever asked, to type the appropriate commands (clearly indicated in boldface) and observe the results before continuing. There will undoubtedly be some frustrating instances where obviously incorrect answers are obtained, or syntax errors are made as a result of minor errors in typing. In these instances, students are encouraged not to lose their patience and remember that all commands must follow the exact syntax rules. It may also help to remember that computers do what their programmers and users tell them to do, which may not be what we want them to do.

Experimentation at Whittier College

As a result of an instrumentation grant from the National Science Foundation for computer-based instruction in mathematics, a math lab was set up in the department of mathematics and computer science in the fall of 1989, and the author was scheduled to teach the calculus sequence using Maple. In addition to the occasional use of computers in the classroom during the lectures, students were required to take part in lab sessions a minimum of one hour per week. The greatest challenge of the course was to select appropriate topics and design lab manuals to cover those topics, explaining the concepts involved and introducing the appropriate Maple commands in an orderly explaining the concepts involved and introducing the appropriate Maple commands in an orderly fashion. At the beginning of each lab session, handouts covering a particular topic were given to pairs of students using a Macintosh. They were required to follow the instructions on the handouts and submit the completed worksheet on that lab session by the end of the week. Since a traditional calculus book was used as a text, and regular homework problems were also assigned, each worksheet usually contained only two or three problems in order not to overburden the students. This text grew out of the revised versions of those handouts and worksheets. The problems on the worksheets can easily be enhanced, should that be desirable by some instructors for their particular class. The topics covered in these labs are arranged in basically the same order as that of the text: *Calculus*, 2nd ed., James Stewart, Pacific Grove, CA: Brooks/Cole Publishing Company, 1991. Exercises marked with an asterisk are obtained from this text.

Nonstandard Functions

In some of the lab sessions, it is necessary to repeat a certain sequence of commands several times. To save the students time, some of these sequences of commands are built into Maple functions, so that the student needs only to call each one by name when needed. Since these are not part of the standard Maple, they have to be included in the Maple folder and "read" before being invoked. A disk containing these functions (in the form of **m.files**) is available[1] to the instructor of the course, for inclusion in the Maple folder within the file server. Note that these m.files must be *individually* placed into the appropriate Maple folder. Instructions on how to read these files are given in the individual lab session, when necessary. For the convenience of the course instructor, the m.file disk also include files containing the full set of Maple commands used in all of the lab sessions. These enable the instructor to observe the result of executing all the commands in a given lab without having to retype them.

Acknowledgments

In March of 1989, I took part in an NSF-sponsored workshop on Computer Algebra Systems at Pomona College. The workshop was under the supervision of Professor Don Small of Colby College. It was a very stimulating and helpful workshop, and was instrumental in developing many of the ideas in some of these lab sessions. I express my deepest appreciation to professor Small for conducting such a well-organized and fruitful workshop. I would also like to thank my many devoted calculus students who worked very seriously during those lab sessions, and encouraged me, through personal contacts and student course evaluations, to continue preparing these notes and bring the project to completion.

Special thanks go to Jeremy Hayhurst, Bob Evans, Nancy Champlin, and Leslie Arthur of Brooks/Cole Publishing Company for their assistance in making this publication possible. Finally, I would like to thank my wife, Janet, and my son, John, for their continued support during the project.

---

[1]Contact the publisher, Brooks/Cole Publishing Company, 511 Forest Lodge Rd., Pacific Grove, CA, 93950, attention Software Support, or call 408-373-0728.

Contents

Lab 1:	Introduction to Maple	1
Lab 2:	Functions	6
Lab 3:	Limits of Functions	10
Lab 4:	The Derivative	13
Lab 5:	Newton's Method For Solving Equations	16
Lab 6:	Graphs of Polynomial Functions	19
Lab 7:	Graphs of Rational Functions	23
Lab 8:	Areas Under Curves and Integrals	27
Lab 9:	Evaluating Integrals by Substitution	31
Lab 10:	Inverses of Functions	35
Lab 11:	Functions Defined by Integrals	39
Lab 12:	Solving Differential Equations	42
Lab 13:	Integration by Parts, Reduction Formulas, and Pattern Recognition	46
Lab 14:	Integrals of Rational Functions	51
Lab 15:	Numerical Integration	54
Lab 16:	Error Bounds in Trapezoidal and Simpson's Rules	57
Lab 17:	Graphs of Parametric Equations	62
Lab 18:	Graphs in Polar Coordinates	64
Lab 19:	Rotation of Conic Sections	66
Lab 20:	Sums and Infinite Series	70
Lab 21:	Taylor Polynomials and Function Evaluation	73
Lab 22:	Vectors in Maple	77
Lab 23:	Plotting Graphs in Three-dimensional Space	81
Lab 24:	Functions of Several Variables (Part A)	85
Lab 25:	Functions of Several Variables (Part B)	89
Appendix A:	Maple Functions Used in This Book	A - 1

1 Introduction to Maple

Throughout these laboratory sessions, it is assumed that you are using a microcomputer, either a Macintosh or IBM-PC (or compatible), with a copy of the **Maple V** program loaded or mounted from a file server. The Maple icon (a maple leaf) should be visible. When you double-click on the Maple application, you will get a Maple V start-up screen and then a blank worksheet. On a Macintosh the • symbol will appear on the screen. In the IBM-PC version of Maple, the prompt is >. The prompt will appear anytime that Maple is ready for a new command. All Maple commands *must terminate with a semicolon* (to see the result) *or a colon* (if you don't want to see the result). You must press the Enter key after the semicolon or the colon. To quit Maple, type quit, or choose **Quit** from the File menu.

Arithmetic

Maple uses the ordinary symbols for arithmetic operations, including the asterisk (*) for multiplication. Type the following commands, and observe the results. (*Warning*: Maple is case-sensitive, meaning that Maple treats the upper- and lower-case alphabetic characters differently. Maple is *not* space-sensitive, which means that hitting the space bar will not affect your input.)

```
2 + 5;
```

Maple will print the answer, 7, on the next line.

```
10 * 25 + 30;
```

Is the answer what you expected?

```
10 + 25 * 30;
```

Notice the order in which the operations are carried out. Multiplication is done before addition.

```
ifactor(");
```

The " symbol tells Maple to substitute the previous expression.

```
ifactor(32340);
```

Experiment with different numbers. Don't try numbers that are more than 20 digits long, because factorization of *some* large numbers is very time consuming.

```
gcd(24, 150);
```

The **gcd** function returns the greatest common divisor of two integers. Try this command with other pairs of numbers. Here you may try arbitrarily large numbers.

```
2 ^ 12;
```

The ^ symbol stands for exponentiation. The ** symbol serves the same purpose. Try both.

```
2/3 + 4/5;
```

Note that the answer is a fraction, not a decimal number. This is one of the important differences between Maple and your calculator. Maple does rational arithmetic as well as decimal expansions. Your calculator generates a floating-point (decimal) approximation to this fraction. Try it now.

```
evalf(");
```

This command evaluates the expression in the parentheses as a real number with a certain number of significant digits. The default is 10 digits.

```
evalf(Pi, 25);
```

This is the approximate value of π with 25 significant digits. In Maple **Pi** stands for π.

```
sin(3);
```

The exact value, as a constant.

```
sin(3.0);
```

The approximate value of sin(3). Precision is 10 digits by default. You can increase the precision, to 50 for example, for all calculations in your Maple session by typing **Digits:= 50**. Try it.

Algebra

Maple is designed to manipulate symbols and algebraic expressions. In accepting the value of an expression, Maple uses a one-dimensional format (that is, everything is typed in one line), but it will print the expressions in normal mathematical format. A whole expression can be assigned a name by the use of the := symbol. Try the following commands:

```
g := x^2 - 3 * x + 2;
```

Maple will echo this definition using normal mathematical format.

```
factor(g);
```

Maple tries to factor the polynomial. The last two commands could be combined into the single command **factor(x^2 – 3 * x + 2);**. Experiment with this command for several other polynomials. In general, Maple is able to factor a polynomial with rational coefficients, if it has any factors with rational coefficients. If unable to factor a polynomial, Maple will just print the whole polynomial.

```
f := (x + 2) ^ 3;
```

Maple will print this in normal mathematical format.

```
expand(f);
```

Maple will expand it into a polynomial in x in standard form.

```
subs( x = 5, f);
```

The expression will be evaluated at $x = 5$. Note that we could not have used the notation $f(5)$, since expressions are not handled the same way as functions. Later we will learn how we could have defined f originally as a function. Then $f(5)$ would make sense and produce the desired result.

```
solve(f = 0, x);
```

This command attempts to give all the solutions of the equation $f = 0$. Since, in this case, the equation has a root $x = -2$ of multiplicity 3, the number -2 is listed three times. For equations of degree less than or equal to 4, Maple will print all real and complex roots. Due to the nature of the algorithms for solving cubic and quartic equations, the solutions will sometimes be in awkward forms. Often it is more convenient to assign a name to the list of roots by the command **r := solve(f = 0, x);**, and then have each of the roots r[1], r[2], … printed in compact form by issuing the command **evalf(r[1])** or **evalc(r[1])**, and so on. The **evalc()** command prints the value of a complex number in the simple form $a + bI$, where a and b are real numbers.

```
solve(f = 1, x);
```

This lists the three roots of the equation $f = 1$. Two of the roots are complex.

```
r := solve(f = 1, x);
```

Assigns the list of roots to the vector r.

```
evalc(r[2]);
```

This is the second root in standard complex form.

```
solve(sin(x) - x + 1 = 0, x);
```

Maple tries, but cannot solve this equation. We will later learn about another Maple command that will be useful in many situations, but here we would like to show how one can approximate the solution(s) by looking at the graph of the expression $\sin(x) - x + 1$. The command for plotting the graph of an expression p depending on x, from $x = a$ to b, is **plot(p, x = a..b);**. If the domain is not specified, Maple assumes it is from -10 to 10. Now try the following:

```
plot(sin(x) - x + 1);
```

Having noticed that there is a root somewhere between 1 and 4, we can use the **plot** command with smaller and smaller ranges for *x* until we get reasonably close to the desired root. We will see later that we can approximate solutions to any desired accuracy.

```
plot(sin(x) - x + 1, x = 1.8 .. 2.2);
plot(sin(x) - x + 1, x = 1.92 .. 1.95);
```

and so on.

Having learned this method of solving equations, we can now introduce the **solve** command. This command, which solves the equations for a floating-point solution, works in many cases and produces an approximate solution quickly. However, it does not necessarily produce all the solutions.

```
fsolve(sin(x) - x + 1 = 0, x);
```

The **solve** command can also be used to solve a given equation for one of the unknowns in terms of the other unknowns. Systems of equations and inequalities can be solved as well. Try the following commands:

```
solve(x^2*y - 2*x + y = 0, y);
```

This solves for *y* in terms of *x*. To solve the same equation for *x* in terms of *y*, you need to replace the last *y* in the command by *x*. (Remember, you can do this by editing the preceding command line rather than typing a whole new line.) The **solve** command can also be used to solve systems of equations. Here the equations and the unknowns must be placed within braces:

```
solve({2*x + 3*y = -1, x - y = 2}, {x,y});
```

Finally, the **solve** command can also be used to solve certain inequalities:

```
solve(abs(2*x - 1) < 3, x);
```

Exercises with Maple

1. Let p be the polynomial $4x^4 + 8x^3 - 13x^2 - 2x + 3$.

a) Evaluate p at $x = 5$, 1.34, and −120.

....

b) Find all the roots of p by using the **solve** command.

....

c) Have Maple factor p, and write the result here. Does this factorization lead to the same roots as in (b)?

....

2. Recall that if $x = a$ is a root of the polynomial p, then $x - a$ must be a factor of p. Find a polynomial of degree 5 with integer coefficients that has −3, 12/23, 5/7, −2/3, and 0.7 as its roots. Explain briefly what Maple commands you used.

....

....

....

3. For each positive integer n, $n!$ (read n-factorial) is defined by: $n! = (1)(2)\ldots n$. Maple understands this symbol and produces its actual value. If you ask Maple to show $n!$ for different values of n, you will note that as n gets larger, the number of trailing zeros in $n!$ also gets larger. Experiment with this and see whether or not you can find a pattern. Can you conjecture a procedure or an explicit formula for finding the number of trailing zeros in $n!$?

....

....

....

2 Functions

Earlier we learned that one can define an expression such as $x^2 + 2x + 3$ in Maple, by

```
f := x^2 + 2*x + 3;
```

Then the name f can be used in subsequent commands to refer to that expression. For example, `solve (f = 0, x);`, or `plot(f, x = -5 .. 5);`, or `factor(f);`. We also learned that to evaluate the expression at $x = a$, we had to use the command `subs(x = a, f);`. It would have been simpler and more in line with ordinary mathematical notation to refer to the evaluation as $f(a)$. But Maple does not understand that notation unless f is defined as a function. To define such a function, we use the Maple command

```
f := proc(x) x^2 + 2*x + 3  end;
```

Maple repeats the definition of the function in the more familiar format and remembers it. Now you can use commands such as

```
f(a);
f(2);
plot(f(x));
```

The word "**end**" is a crucial part of the definition of a function. Anything typed before that word is considered a part of the definition of the function, and forgetting to include it can lead to frustration. Maple V has a simpler method (than earlier releases) for defining functions. It takes the form

```
f : = x -> x^2 + 2*x + 3;
```

We will see later that many other Maple commands, such as **diff**, **int**, and **taylor**, may use a function f as an argument. Step functions can also be defined using this method. For example, to define the function

$$g(x) = \begin{cases} 2x + 1 & x \le 0 \\ x^2 + 1 & x > 0, \end{cases}$$

we use the Maple command

```
g := proc(x) if x <= 0 then 2*x + 1 else x^2 + 1 fi end;
```

Don't forget the keywords **fi** and **end**! Now try

```
g(a);
```

Note that Maple prints an error message and does not produce the value $g(a)$, unlike the case for f. This is because a is unknown. Hence Maple cannot tell whether it is positive or negative and therefore cannot come up with an expression for $g(a)$. However, Maple has understood and will remember the definition of g for the remainder of the session, and can evaluate it for a specific value of x. To verify this, try

```
g(-2);
g(5);
plot(g,   x = -7 .. 5);
```

Multistep functions can be defined in a similar manner. For example, to define the function h given by

$$h(x) = \begin{cases} x+3 & x < 2 \\ 5 - x^2 & -2 \le x \le 2 \\ 3 - x & x > 2, \end{cases}$$

we proceed as follows:

```
h := proc(x)   if x < -2   then   x + 3
                    elif   x <= 2   then    5 - x^2
                    else 3 - x
                    fi
         end;
```

The whole command could be typed on one line. Here it is separated into five lines for clarity. The Return key can be used to type a long command into several lines.To enter a multiline command in Maple V, simply place the insertion point anywhere in the command and press the Enter key. However, for earlier versions of Maple, a multiline command is entered by placing the insertion point in the beginning of the command, dragging all the way to the end, and then pressing the Enter key.

Try evaluating this function at a few points of your choice. Then sketch its graph, by using the command

```
plot  (h);
```

Now try

```
plot({f,  g},  x = -5 .. 5);
```

This is how more than one graph can be plotted on the same set of axes. In fact, this feature of Maple can be used to see the effect of certain simple transformations on the graph of a function. The following commands perform horizontal shifting, vertical shifting, horizontal scaling, and positive and negative scaling, respectively:

```
plot({f(x),   f(x - 3)});
plot({f(x),   f(x) - 50});
plot({f(x),   f(2 * x) });
plot({f(x),   2 * f(x) });
plot({f(x),  -2 * f(x) });
```

As another example of a transformation on the graph of a function, consider the graph of the fuction g, where $g(x) = |f(x)|$. It is clear from this relation that the graph of g will never enter below the x-axis. In fact, the graph of g coincides with that of f except on those intervals over which $f(x) < 0$, in which case the graph of g is simply the reflection of the graph of f about the x-axis. Here is a specific example:

```
f := proc(x)  4 - x^2  end;
g := proc(x)  abs(4 - x^2) end;
plot( {f , g}, x = -5 .. 5);
```

Exercises with Maple

1. Recall that $n! = (1)\ (2)\ (3)\ \ldots\ (n)$. Have Maple draw the graph of $f(x) = \dfrac{x^n}{n!}$ for $n = 1, 2, \ldots, 6$.

a) How are these graphs different for the even values of n versus the odd values of n? Can you justify your answer by considering the definition of the function?

....

....

b) What happens to these graphs as n gets larger?

....

....

2. Enter the Maple command

```
f := proc(x) if x < -2 then -2-x elif x < 2 then sqrt(4-x^2)
else x-2 fi end;
```

a) Write the definition of this function in the usual mathematical notation. Have Maple plot the graph of *f*. Is the graph as you expected?

....

....

b) Define a new function *g* by `g := proc(x) f(x - 1) end;`

Now have Maple plot the graph of *f* and *g* on the same set of axes. Then explain how the graph of *g* is related to that of *f*.

....

c) What Maple command would you use to define a function *h* obtained from *f* by shifting it three units to the left? Try plotting *f* and *h* on the same axes to see whether or not you get the desired result.

....

d) How would you define the function *k* whose graph is the reflection of the graph of *f* about the *x*-axis? Try plotting *f* and *k* on the same axes to verify that you have found the right function *k*.

....

3*

a) Starting with the function $|x|$ and applying appropriate transformations, decide what the graph of $f(x) = ||\,|x| - 2| - 1|$ should look like.

....

b) Have Maple draw the graph of $f(x)$ on $[-4, 4]$.

3 Limits of Functions

The purpose of this lab session is to help understand the concept of the limit of a function. Intuitively, to say that the function f has a limit of L as x approaches a means that as x gets closer and closer to a, $f(x)$ gets closer and closer to L. This concept can be formalized to give a more precise definition (the so-called epsilon-delta definition, as discussed in your calculus text).

Maple has a built-in function, **limit**, that allows you to find the limit of a function at a point, if it exists. The syntax is `limit(f(x), x = a);`, where a is a specified real number. However, our purpose here is to clarify the foregoing intuitive definition by actually printing out a table of values of $f(x)$ for those values of x that are very close to a both from left and right. We will also see that plotting the graph of a function on a shrinking set of intervals about the point a can help explain the behavior of f near that point and determine whether or not the limit exists.

To reduce the amount of typing and to avoid syntax errors, the commands directing Maple to print out a table of values for x and $f(x)$ are written into a Maple procedure. To include this procedure in your Maple session, you must use the **read** command (we will see a complete example shortly). This command loads in the procedure from a Maple library, and if you define the function f (using the method we learned in Lab 2), you are ready to have the table of values printed. The table will have three columns, labeled h, $f(a - h)$, and $f(a + h)$. h takes on values $\frac{1}{10}$, $\frac{1}{100}$, $\frac{1}{1000}$, and so on. The next two columns show what happens to $f(x)$ when x gets closer and closer to a from left and right. We are now ready to look at an example:

```
read( `LimTable.m`);
```

This makes the procedure **LimTable** available to this Maple session. Note that the delimiters are back quotes, not apostrophes.

```
f := proc(x)   sin(x)/x   end;
```

This defines the function $f(x) = \dfrac{\sin x}{x}$.

```
LimTable(f, 0);
```

This asks that a limit table for the function f near the point $a = 0$ be printed. Inspect the second and third columns of this table to see whether you can conclude that the function has a limit as x approaches 0. To see that your conclusion agrees with Maple's limit-calculating routine, type

```
limit (sin(x)/x , x = 0);
```

To check whether you could have gotten the same information from the graph of f, plot the graph of f over several intervals of decreasing size, and study the graphs.

```
plot (f );
```

By default, this assumes the interval [−10, 10] for x. Now try it for shorter and shorter intervals about zero:

```
plot (f, x = -1 .. 1);
plot (f, x= -.1 .. .1, y = .995 .. 1.005);
```

Now define a new function f, as follows:

```
f := proc(x)  if  x < 0  then  2 - x  else  x + 1  fi  end;
```

To have the limit table printed for this function, place the cursor anywhere on the line containing the **LimTable** command and press the enter key. Does the limit of f exist as x approaches 0? Verify your answer by having Maple plot the graph of f on shrinking intervals about 0. Since f is defined as a step function, the command `plot(f(x), x = -5 .. 5);` will not work. You must use

```
plot(f, x = -5 .. 5);
```

What happens when you try to use Maple's limit command to find the limit of this function?

The **LimTable** procedure can also be used to examine the values of $f(x)$ as x approaches $+\infty$, or $-\infty$. To do this, use the words "infinity" or "–infinity" in place of a. Try the following:

```
h := proc(x)  (x^2 - 1)/(2*x^2 - 9*x + 8)  end;
LimTable(h , infinity);
LimTable(h , -infinity);
```

Examining the table of values is helpful in understanding the concept of limit. In practice, various limit theorems are used to determine the limit of a function. Furthermore, in certain special cases, the limit table could lead to incorrect conclusions. For example, consider the function

$$f(x) = \sin \frac{\pi}{x}$$

```
f := proc(x)  sin(Pi/x)  end;
LimTable(f , 0);
```

It appears from this table that the limit is zero. However, it can be shown that $\lim_{x \to 0} \sin(\pi/x)$ does not exist. In fact, $\sin(\pi/x)$ can be made to approach any number between −1 and 1 by allowing x to approach 0 along an appropriate sequence.

In using the **LimTable** procedure, if the numbers involved are too small, they are printed in the Float notation, which may be unfamiliar to you. In such cases, you may use the **limtab** procedure instead. You will encounter this problem when you do Exercise 2 of the following set.

Exercises with Maple

1. Enter the function $f(x) = \dfrac{x^2 - 3x + 2}{x^2 - 2x}$.

a) Use the procedure **LimTable** to observe the table of values of f near the point $x = 2$. Round off the last values on the 2nd and 3rd columns to 10 decimal places, and enter them here. Does it appear that f has a limit as x approaches 2? If so, what is it?

....

b) Have Maple plot the graph of f on intervals of length 1 and 0.1 with center at 2. Do the graphs seem to confirm your answer to (a)?

....

c) What answer do you get when you apply the limit command to f at 2?

....

d) Find $\lim f(x)$ as $x \to 2$, using appropriate limit theorems.

....

2. Enter the function $f(x) = x \sin \dfrac{1}{x}$, and repeat (a) through (d) of Exercise 1 for this function and the point $x = 0$ (instead of $x = 2$).

....

3* Have the procedure **LimTable** estimate the value of the function $f(x) = (1 + x)^{1/x}$ near the point $x = 0$. (It will be shown later that $\lim\limits_{x \to 0} (1 + x)^{1/x}$ is the number e, the base of the natural logarithm.)

4 The Derivative

The derivative of the function f at a is defined by

$$f'(a) = \lim_{h \to 0} \frac{f(a+h) - f(a)}{h}.$$

Even though shortcut methods for calculating the derivatives of various functions can be found in your calculus text, the foregoing definition is absolutely essential, and all the theorems and formulas concerning the derivative are based on this definition. In this lab we will use Maple's computational capability to directly calculate the derivative of a function using the foregoing definition. Let us start with a simple function

```
f := proc(x) x^2 - 5 * x end;
```

We will divide the application of the definition into several steps. Try the commands

```
f(1 + h);
f(1 + h) - f(1);          or simply " – f(1);
(f(1 + h) - f(1))/h;      or simply "/h
simplify(");
limit(" , h = 0);
```

You should get a numerical answer of –3, which is in fact the value of $f'(1)$. By replacing all occurrences of the number 1 by the letter x in the foregoing sequence of commands, we can determine the derivative of this function at the unspecified point x. You may now make these alterations. The final result should be $2x - 5$.

Maple has a built-in command that determines the derivative of f at a given point, as well as the derivative at an unspecified point x. To verify the answer we obtained using the definition, try the commands

```
diff(f(x) , x);           The answer should be 2x – 5.
subs(x = 1, ");           The answer should be –3.
```

Now repeat the steps of the definition with the function

```
f := proc(x)   (x^2  - 3)/(2*x + 5)    end;
```

and also with the function. (Here, skip the **simplify** command.)

```
f := proc(x)   sin(5*x)   end;
```

To allow for more practice with the steps of the definition of the derivative and to reduce the amount of typing, again a Maple procedure containing these steps is created. To include this procedure in your Maple session, enter the command

```
read (`Der.m`);
```

Now, assuming that you have already defined your function *f*, to get the derivative of *f* at 1, use the command

```
Der(f , 1);
```

On the other hand, if you want the derivative at an unspecified point *x*, then use the command

```
Der(f);
```

Of course, in the latter case your final result will be a function of *x*. Repeat this with two other functions of your choice. Make sure to compare your results to what you would get if you applied the **diff** command directly.

As mentioned earlier, our main objective in this lab is to emphasize the definition of the derivative. In application problems, where the derivative of a function is needed, we will frequently use the **diff** command to obtain the derivative. For example, if the cost of producing *x* items is given by the function

$$C(x) = -0.02\,x^2 + 25x + 3\sqrt{x} + 1200,$$

then the marginal cost at the production level of $x = 200$ can be obtained by

```
diff(-0.02 * x^2 + 25*x + 3 * x^(1/2) + 1200, x);subs(x = 200, ");
```

Exercises with Maple

1. Using the definition of the derivative and following the instructions given earlier, determine the derivative of $f(x) = x^3 + 1$ at x. Compare the final result with the result of the direct application of the **diff** command. Are they the same?

....

2. Define the function f by

$$f(x) = \begin{cases} x \sin\dfrac{1}{x} & \text{if } x \neq 0 \\ 0 & \text{if } x = 0 \end{cases}$$

a) Use Maple's **diff** command to find out whether the derivative of f at 0 exists.

b) Use the definition of the derivative or the **Der** procedure defined in this lab to find the derivative of f at 0. Do you see the problem?

....

3* Repeat Exercise 2 with the function

$$f(x) = \begin{cases} x^2 \sin\dfrac{1}{x} & \text{if } x \neq 0 \\ 0 & \text{if } x = 0 \end{cases}$$

True or false: In finding the derivative of a function f at $x = a$, we can always first find the derivative at x, then replace x by a. Explain.

....

....

....

5 Newton's Method for Solving Equations

We saw earlier that except for polynomial equations of degree four or less and certain other special cases, most equations are extremely difficult, if not impossible, to solve precisely. We also saw that the **fsolve** command can usually find a solution for most equations, provided that one exists. In this lab we will learn a technique for approximating the solution of the equation $f(x) = 0$. This technique, known as Newton's Method, works as follows

> Choose an initial value x_0, and then find a sequence of numbers $x_1, x_2, \ldots,$ $x_n, \ldots,$ where x_{n+1} is the x-intercept of the tangent line to the graph of $y = f(x)$ at x_n. A simple calculation shows that we must have
>
> $$x_{n+1} = x_n - \frac{f(x_n)}{f'(x_n)}.$$

To guarantee that the equation has a solution, we will consider f over an interval $[a, b]$, where $f(a)$ and $f(b)$ have opposite signs. Then, assuming that f is continuous on $[a, b]$, by the Intermediate Value Theorem there must exist a number c between a and b such that $f(c) = 0$. This will guarantee the existence of a solution on $[a, b]$, but Newton's method may still fail to find an approximation to the solution. It turns out that there are several other conditions that will collectively guarantee that Newton's method will succeed. These conditions can be found in a text on numerical analysis.[1]

We can have Maple do all the tedious calculations and print the values of the x_n's for as many iterations as we please. Again, to reduce the amount of typing, the commands to do this are included in a procedure. The procedure needs to have four arguments: the function f, the two endpoints of the interval $[a, b]$, and the number of iterations you wish to have. Note that if $f(a)$ and $f(b)$ do *not* have opposite signs, you will *not* be able to use this procedure. You can use the **read** command to include this procedure in your Maple session:

```
read `Newton.m`;
```

Now, as a first example, let's use Newton's method to approximate the positive square root of 5, that is, a solution of the equation $x^2 - 5 = 0$:

```
f := proc(x)   x^2 - 5  end;
```

For the interval $[a, b]$, we could choose [2, 3]. Now type the command

```
Newton(f, 2, 3, 10);
```

This example shows how powerful Newton's method can

1 See Peter Henrici, Elements of Numerical Analysis, New York: Wiley, 1969, p. 79.

be. After just three iterations, the solution is correct within ten significant digits. Now try for a solution to the equation $x^3 - 3x + 1 = 0$:

```
f := proc(x)   x^3 - 3*x + 1   end;
```

Since $f(1) < 0$ and $f(2) > 0$, we choose $a = 1$ and $b = 2$:

```
Newton(f, 1, 2, 10);
```

Again, it looks as if a solution to the equation has been found that is correct to 9 decimal places after only three iterations. To verify that this number is in fact an approximation to a solution, try

```
f(1.532088886);
```

Maple confirms this by evaluating it to zero. Sometimes it may take many iterations for Newton's method to finally produce an approximation to the root. A case in point is the function

```
f := proc(x)   x^10 - 1   end;
```

It takes about 86 iterations to get the obvious solution $x = 1$, if we choose the interval $[-0.5, 1.1]$:

```
Newton(f, -0.5, 1.1, 90);
```

Newton's method may not produce a solution at all. Try the function $f(x) = \sqrt[3]{x}$ on the interval $[-2, 4]$:

```
f := proc(x)   x ^ (1/3)   end;

Newton(f, -2 , 4 , 30);
```

Notice that each estimate gets larger than the previous one in absolute value; hence $\{x_n\}$ does not converge to a finite number, even though the obvious root $x = 0$ lies in the foregoing interval. Note that in both of these examples, the interval under consideration includes the point $x = 0$, which is a critical point of the function.

Procedure **Newton** uses the midpoint of the original interval as the initial estimate of the root. In general, the behavior of Newton's method may depend greatly on the choice of this initial estimate. For example, consider the function

$$f(x) = x^3 - x - 1$$

on two intervals centered at two close-by points, 0.600 and 0.577. The following commands will show that in the first case the root is determined after about 12 iterations, whereas in the second case it will take about 60 iterations to arrive at the root (approximated to 10 significant digits):

```
f := proc(x)   x^3 - x - 1 end;
Newton(f,  -0.4 , 1.6 , 20);
Newton(f,  -.423 , 1.577 , 65);
```

Exercises with Maple

1. Use the procedure **Newton** to find the 7th root of 100. What function did you use?

....

2. Let $f(x) = x^5 - 4x^2 + 3x - 2$.

a) Verify that f has a root on the interval [1, 2].
b) Use the procedure **Newton** to find an approximation to this root. How many iterations did you need in order to get the root with 9 digits of accuracy ?

....

c) Find the root using the **solve** or the **fsolve** command. Is the answer the same?

....

3. Repeat (a) through (c) of Exercise 2 for the function $f(x) = \sin x - x + 1$.

....

4* a) By drawing the graphs of $f(x) = x$ and $g(x) = \tan x$ on the same set of axes, find the approximate value of the root of the equation $\tan x = x$.

....

b) Use the procedure **Newton** to find the approximation of the root using the interval [4.3, 4.6].

....

6 Graphs of Polynomial Functions

In this lab we will use a systematic method to arrive at the information that can be used in sketching the graph of polynomial functions. This method can also apply, in theory, to other functions, but the equations and the inequalities involved in the case of polynomial functions are solved easily by Maple, especially if the polynomials obtained by differentiation happen to have rational roots. We will go through the procedure with a simple example first. You will need a pencil and some paper to record the information as we go along. Start by defining the function *f*:

```
f := proc(x)   x^3 - 3*x +  2  end;
```

Step 1: Have Maple calculate $f'(x)$, find the roots of $f'(x)$, that is, all the critical points, and determine the sign of $f'(x)$.

```
f1  := diff  (f(x),x);
```

Maple gives the first derivative of *f*.

```
solve(f1  =  0);
```

Maple gives the list of roots, including the repeated roots, if they are all rational.

```
solve(f1  >=  0);
```

This gives the intervals over which $f'(x)$ is positive; hence *f* is increasing. On the rest of the real line, $f'(x)$ would be negative; hence *f* will be decreasing. Write this information on your paper.

Step 2: Have Maple calculate $f''(x)$, find the roots of $f''(x)$, that is, possible inflection points, and determine the sign of $f''(x)$.

```
f2 := diff  (f1  ,  x);
```

Maple gives the second derivative of *f*.

```
solve  (f2  =  0);
```

This gives possible inflection points.

```
solve  (f2 >=  0);
```

This gives the interval over which $f''(x)$ is positive, thus determining where the graph of *f* is concave up. On the remaining part of the real line, the graph will be concave down. This also determines whether the point(s) found by the previous command lead to inflection points (if the concavity changes). Write this information down as well. So far, we have the following information:

a) f has critical points at $x = -1$ and $x = 1$.
b) f is increasing on $(-\infty, -1)$ and on $(1, \infty)$, and it is decreasing on $(-1, 1)$.

It follows from (a) and (b) that f has a relative maximum at $x = -1$ and a relative minimum at $x = 1$.

c) The graph of f is concave up on $(0, \infty)$; concave down on $(-\infty, 0)$.
d) The graph of f has an inflection point at $x = 0$.

Step 3: Evaluate f at a few points (including the critical and inflection points), plot these points, and connect them using a continuous curve, keeping the foregoing properties in mind. You should also find the x- and y-intercepts. Do Step 3 now.

Now compare your sketch with Maple's graph obtained by the command

```
plot(f);
```

Note that Maple's graph does not look like yours. Because of the different scales on the x- and y-axes, part of the graph is distorted. To see a more realistic version of the graph drawn by Maple, restrict its domain:

```
plot(f, x = -2.5 .. 2.5);
```

This should now be quite similar to the graph you drew by hand.

Important note: If the polynomials $f1$ and $f2$ have irrational or complex roots, the **solve** command may be unable to find the roots and may print the solutions in the **RootOf(...)** format, or, in the case of cubic and quartic polynomials, it may produce pages of "stuff" on the screen. You can just ignore all of that and use the **fsolve** command instead. This command will find all the real roots in approximate form and print them as a list. It will print multiple roots only once, however. This next example is a case in point:

```
f := proc(x)  x^5 - 4 * x^4 - x^3  + 5 * x^2 - 1  end;
f1 := diff (f(x), x);
```

If you now try **solve** ($f1 = 0$), you will not get anything useful. So try

```
fsolve(f1 = 0);
```

Maple finds the approximate values of the four roots. They are: – 0.8, 0, 0.8, and 3.2 (rounded to one decimal place). Solving the inequality **solve (f1 >= 0);** will not work, but since we know all the roots of $f1$, we can determine its sign by dividing the real line into five segments using the roots as the cutoff points, evaluating $f1$ at a well-chosen point in each segment, and observing the sign:

```
subs (x = -1,   f1);
subs (x = -.5,  f1);
subs (x = .5,   f1);
subs (x = 1,    f1);
subs (x = 4,    f1);
```

Since this is positive, $f'(x) > 0$ on $(-\infty, -8)$.
Since this is negative, $f'(x) < 0$ on $(-8, 0)$.
Since this is positive, $f'(x) > 0$ on $(0, .8)$.
Since this is negative, $f'(x) < 0$ on $(.8, 3.2)$.
Since this is positive, $f'(x) > 0$ on $(3.2, \infty)$.

This determines the intervals over which f is increasing or decreasing. It also implies that f has a relative maximum at $x = -.8$ and $x = .8$, and that it has a relative minimum at $x = 0$ and $x = 3.2$. This completes Step 1. Now we do Step 2:

```
f2 := diff (f1, x);
fsolve  (f2 = 0);
```

Maple shows the approximate roots. To one decimal place, they are –0.5, 0.5, and 2.4. These numbers divide the real line into four segments, and, just as before, you can determine the sign of $f2$ by evaluating $f2$ at a well-chosen point inside each segment. This will show where the graph of f is concave up or down, and will also determine the inflection points. It turns out that all three roots of $f2$ are, in fact, inflection points. Now evaluate f at the critical and inflection points, and sketch the graph on the paper. Let's see what Maple's graph looks like:

```
plot(f);
```

Because of scaling, the graph looks distorted again. Let us try

```
plot(f, x = -2 .. 4);
```

This looks a lot better, and closer to your graph on the paper. You may plot the graph on even smaller intervals to see its behavior near the critical points.

Exercises with Maple

1* a) Follow the procedure described in Lab 6 to get a sketch of the graph of $f(x) = x^4 - 3x^3 + 3x^2 - x$ on the paper. List the intermediate results here.

....

b) Have Maple plot the graph, and then compare. Which domain gives the least distorted view of the graph?

....

2. Repeat Exercise 1 for the function $f(x) = x^6 - x^2 + 1$.

....

....

3. Do the same for the function $f(x) = 4x^5 - 20x^4 + 17x^3 + 4x + 1$.

....

....

7 Graphs of Rational Functions

Although Maple's **plot** command may work fine with very simple rational functions, in general it does not work properly with most rational functions. This is mainly due to the fact that the absolute value of the function becomes very large near the points where the vertical asymptotes occur. In this lab we will learn how to use Maple's calculational capability to obtain useful information and draw an approximate sketch of the graph of a rational function in a systematic way. A classroom discussion of this topic appears in an interesting article by professor Don Small of Colby College.[1]

Suppose we want to sketch the graph of:

$$f(x) = \frac{x^2 - 2x}{3x^4 - 28x^3 + 84x^2 - 96x + 35}.$$

First we define the function f

```
f := proc(x) (x^2 - 2*x)/(3*x^4 - 28*x^3 + 84*x^2 - 96*x + 35)
end;
```

Maple's **plot** command will take some time and, unless supplied with a carefully chosen range of values for x and y, will not reveal much about the graph. Instead, we will carry out the following steps and record all the information on a piece of paper:

Step 1: Find all the critical points by finding the derivative and setting its numerator equal to zero:

```
d1 := diff( f(x) , x);
simplify(");
ex := numer(");
fsolve ( ex = 0 , x );
```

So f has three critical points: – 0.80, 1, and 3.54 (all rounded to two decimal places). Now find the relative extrema using the Second Derivative Test:

```
d2 := diff( d1 , x);
subs( x = -.80 , d2);
```

Since this number is negative, f has a relative maximum at $x = -0.80$. Some caution is necessary here if this number is very close to zero. Since we are using only an approximation to the actual critical number, it is conceivable that f'' could

[1] Don Small, "Script for a CAS Classroom Exercise," Computer Algebra Systems in Education NEWSLETTER, Dept. of Mathematics, Colby College, No. 5, Jan. 1989, pp. 5 - 7.

have been precisely equal to zero at the actual critical number, in which case the Second Derivative Test cannot apply. Now the next critical point:

```
subs(x = 1 , d2);
subs(x = 3.54 , d2);
```

So f has a relative minimum at $x = 1$.
And f has a relative maximum at $x = 3.54$.

Step 2: Reduce the rational function to lowest terms and find the roots of the denominator to determine the vertical asymptotes. Note that the **normal** command can be used to put a fraction in reduced form.

```
den := denom (f(x));
fsolve (den = 0 , x);
```

In the rare cases when the denominator has only rational roots, the exact roots can be found by using the **factor** or **solve** command instead of the **fsolve** command.

This shows that the graph of f has four vertical asymptotes: $x = .70$, $x = 1.40$, $x = 2.48$, and $x = 4.74$ (again rounded to two decimal places). At this point you should draw the coordinate axes, as well as these vertical lines, on the paper.

Step 3: Determine the horizontal asymptote of the graph of f by finding $\lim f(x)$ as x tends to infinity. If the degree of the numerator is less than that of the denominator, as in this example, this will always be $y = 0$. If the degrees are equal, this would be $y = a$, where a is the ratio of the coefficients of the highest power of x in the numerator and the denominator. Alternatively, you could use the command **limit(f, x = infinity);** or **limit (f, x = – infinity);**

Step 4: Find all the x-intercepts of f by finding all the roots of the numerator. In this example, the x-intercepts are clearly $x = 0$ and $x = 2$. In the general case, you can find the x-intercepts by

```
fsolve(numer(f(x)) = 0 , x);
```

Step 5: Use the roots of the denominator, obtained in Step 2, to express the domain of the function as the union of several open intervals. In this case we obtain the intervals $(-\infty, .70)$, $(.70, 1.40)$, $(1.40, 2.48)$, $(2.48, 4.74)$, and $(4.74, \infty)$. Knowing that the graph of the function is continuous on each of these intervals, and having determined the relative extrema and the x-intercepts, we can determine the approximate shape of the graph over each interval. By evaluating f near the endpoints of these intervals, we can find out on which side of the asymptotes to draw the graph. For the first interval $(-\infty, .70)$, the command

```
f(-100);
```

determines that the graph starts above the horizontal asymptote. On the other hand, the command

```
f(.69);
```

reveals that the curve approaches the vertical asymptote $x = .70$ on the lower end. Knowing that there is a relative maximum at $x = -0.8$ and an x-intercept at $x = 0$, we can basically complete the graph on the first interval. To determine the shape of the graph on the interval $(.70, 1.40)$, we again evaluate f at points near the endpoints:

```
f(0.71);
f(1.39);
```

Since both of these numbers are positive, the second branch of the graph starts at the upper part of the vertical asymptote $x = .70$ and ends up at the upper end of the vertical asymptote $x = 1.40$. This agrees with the earlier information that f has a relative minimum at $x = 1$. Continuing in this fashion, you can now sketch the graph of f on the three remaining intervals. For a more accurate sketch, you can evaluate f at several other points.

Exercises with Maple

1. Apply the systematic method discussed in Lab 7 to sketch the graph of each of the following rational functions.

a)* $f(x) = \dfrac{1 - x^2}{x^3}$.

b) $f(x) = \dfrac{x^3 + x^2}{x^5 - 3x^4 - x^3 + 11x^2 - 12x + 4}$.

Clearly identify all vertical and horizontal asymptotes and all relative extrema.

8 Areas under Curves and Integrals

If the graph of the function f on the interval $[a, b]$ lies above the x-axis, then the area of the region bounded above by the graph, below by the x-axis, and on the two sides by the vertical lines $x = a$ and $x = b$ can be approximated as follows.

Divide the interval $[a, b]$ into n subintervals of equal length, and on each subinterval form a rectangle of height $f(c)$, where c is the left endpoint of the subinterval. Thus if

$$a = x_0, x_1, x_2, \ldots, x_{n-1}, x_n = b$$

are the dividing points, then each subinterval has a width of $\Delta x = \dfrac{b-a}{n}$, and the height of the rectangle built on the kth subinterval is $f(x_{k-1})$. Hence the area of the kth rectangle should be equal to $f(x_{k-1})\ \Delta x$. The sum of the areas of these rectangles gives an estimate of the desired area. But this is just the sum

$$\sum_{k=1}^{n} f(x_{k-1})\Delta x.$$

Since we have $x_{k-1} = a + (k-1)\ \Delta x$, this sum now becomes

$$\sum_{k=1}^{n} f\,[a + (k-1)\Delta x]\ \Delta x.$$

Maple has a built-in function **sum** that can evaluate expressions such as this. For example, to evaluate an expression such as

$$\sum_{k=1}^{n} g(k),$$

we can use the command **sum (g(k), k = 1 .. n)**. Let's do this for a specific function. Suppose we want to find the area of the region bounded above by the graph of $f(x) = x^2 + 1$, below by the x-axis, and on the sides by the lines $x = 0$, and $x = 1$. We proceed as follows:

```
f := proc(x) x^2 + 1 end;
delta := 1/n;
z := sum (f ((k-1) * delta) * delta, k = 1..n);
```

Note that a was replaced by 0.

```
simplify(z);
```

Now by substituting larger and larger values for *n*, we can find better and better estimates for the area. Try

```
evalf (subs(n = 20,  z));
evalf (subs(n = 40,  z));
evalf (subs(n = 100, z));
evalf (subs(n = 500, z));
```

These numbers will be approaching the actual area. In fact, the actual area is just the limit of *z* as *n* approaches infinity

```
limit (z, n = infinity);
```

All of the foregoing calculations could have been done using the right sums instead of the left; that is, for each subinterval, we could have taken the height of the rectangle to be the value of the function at the right endpoint of the subinterval. This amounts to changing $f(k-1)$ to $f(k)$ in the original summation.

When *f* is a polynomial, as in the previous example, Maple can evaluate the sum as a simple function of *n*. But in most other cases, this may not be possible. However, one can always use a specific value in place of *n* in the expressions for **delta** and **z**, thus obtaining a numerical result in each case.

In order to see how the left and the right sums change with the increasing value of *n*, and to reduce repetitious typing of the above commands, several procedures have been provided to carry out the above steps. To include these in your Maple session, use the following command

```
read (`Sums.m`);
```

Now assuming that the function *f* has already been defined, the command **LeftSum (f, a, b, n)** can be used to find the value of the left sum. Here *a* and *b* are the endpoints of the interval under consideration, and *n* is the number of subdivisions. To approximate the area using the value of the function at the right endpoint of each subinterval as the height of the rectangle, use the command: `RightSum(f, a, b, n)`. Finally, the command `SumTable (f, a, b)` will produce a table of values for the left and right sums, starting with $n = 10$ and doubling it until $n = 160$. Now let us try these commands for a new function

```
f := proc(x) x^3 + 4 * x + 3 end;
LeftSum (f, 1, 4, n);
simplify(");
```

Since this is a rational function of n, we can easily find the limit as n approaches infinity:

```
limit (" , n = infinity);
```

Now do the right sum:

```
RightSum (f, 1, 4, n);
simplify(");
limit (" , n = infinity);
```

As expected, both left and right sums have the same limit. Finally, let us take a look at the table of values for the left and right sums for various values of n:

```
SumTable (f,1,4);
```

Important Note: The preceding sums give an estimate of the area under the graph of f only when the graph of f lies entirely above the x-axis.

The purpose of this lab session was to show how we can estimate the area under a curve by adding the areas of rectangles, and to motivate the definition of a definite integral. For functions whose antiderivatives can be found easily, the Fundamental Theorem of Calculus gives the exact value of the area with a lot less calculation. Even if no antiderivative could be found, there are much more efficient methods for approximating the area. We will discuss these methods in Labs 15 and 16.

Maple has built-in commands that can find the definite and indefinite integral of functions. These are:

```
int (f(x), x);
```

This produces the indefinite integral (antiderivative) of the function f. If Maple cannot find an antiderivative, it will echo the command.

```
int (f(x), x = 1..4);
```

This produces the exact value of the definite integral, if Maple is able to find it using various integrating techniques. Otherwise, Maple will echo the command again. But in this case, if an approximate value for the definite integral is desired, one can use the command

```
evalf (");
```

Use these commands to find the following definite integrals:

$$\int_0^3 \sqrt{9-x^2}\ dx, \text{ and } \int_1^2 \sqrt[3]{x^2+1}\ dx.$$

Exercises with Maple

1*
a) Define the function f by $f(x) = x^3 + 2x^2 + x$, on the interval [0, 1].
b) Follow the instructions in Lab 8 to calculate the left sum z.
c) Evaluate z for n = 10, 50, and 100.
d) Use the procedure **SumTable** to print out a table of values for the left and right sums.
e) Find the limit of z as n approaches infinity.
f) Use the **int** command to find the area.
g) Summarize your observations.

....

....

....

....

....

2. a) Use the **SumTable** procedure to see approximate values for

$$\int_0^2 \sqrt{4-x^2}\ dx\ .$$

b) Use Maple's **int** command to find the exact value of this integral, and compare the answer to that obtained in (a).

....

9 Evaluating Integrals by Substitution

The purpose of this Lab Session is to learn how to use Maple to evaluate integrals by the method of substitution. Recall that Maple's **int** command can be used to find the definite and indefinite integral of a function *f*. For example, we can use the following command to find the indefinite integral of the function $f(x) = x\sqrt{9-x^2}$:

```
int(x * (9 - x^2) ^(1/2), x);
```

Maple immediately provides us with the answer. We can easily verify this by differentiating the latter function:

```
diff(" , x);
```

Now if we change x^2 to x^4 in the foregoing function and ask Maple to find the integral again, early releases of Maple will simply display the integral. Maple V will attempt to evaluate this integral, but the result contains complex numbers and another integral of the same degree of difficulty.

```
int(x * (9 - x^4) ^(1/2), x);
```

This does not necessarily mean that the integral cannot be evaluated. In fact, as you may have guessed already, a simple substitution will work in this case. Maple can do the routine calculations in the process of substitution. To do this, we need to load a Maple package called **Student**. This package contains several useful procedures, including **changevar** which is what we need here. To load this package, type

```
with(student);
```

Maple prints the list of available procedures in the package. Using the procedure **changevar**, we can make an appropriate substitution in the indefinite integral and see whether it results in a simpler integral. Type

```
changevar( x^2 = u, Int(x * (9 - x^4) ^(1/2), x), u);
```

Make sure that you type the keyword **Int** with a capital "I". (The function **Int** is a place holder for the function **int**. This means that Maple treats **Int()** like a symbol and does not evaluate it.) This integral looks a lot easier than the original one. As we will see later, it can be evaluated by the substitution $u = 3 \sin t$. But let us see if Maple can do it directly:

```
int (1/2 * (9 - u^2) ^(1/2), u);
```

Note that Maple does not add the arbitrary constant; you will have to do that yourself. To get the indefinite integral as a function of the original variable x, we substitute u in terms of x, using the previously defined relationship between these two variables:

```
subs( u = x^2, ");
```

To verify that this is the right answer, we will differentiate it:

```
diff(",x);
```

We can now simplify this to get the original integrand:

```
simplify(");
```

As another example, we will look at a definite integral. Suppose we want to evaluate the definite integral

$$\int_{-1}^{1} \frac{x^5\,dx}{\sqrt{x^3+9}}$$

We will check first to see whether Maple can find it directly:

```
int(x^5/(x^3 + 9)^(1/2), x = -1 .. 1);
```

If Maple displays the integral, Maple informs us that it is not able to find the exact value of the integral.[1] If it isn't found directly, we will try a suitable substitution:

```
changevar(x^3 + 9 = u, Int( x^5/(x^3 + 9) ^(1/2), x = -1..1),u);
```

This looks easier. See if Maple can do this one:

```
int (1/3 * (-9 + u) * u^(-1/2), u = 8..10);
```

And it gives the exact value of the definite integral. To see an approximation of this value, we can apply the **evalf()** function

---

1 Maple V is able to evaluate this integral directly. Previous releases cannot.

```
evalf(");
```

By the way, if we were interested only in the approximate value of the definite integral, we could avoid all the foregoing work and let Maple use its built-in numerical integration package, by issuing the command

```
evalf( int(x^5/(x^3 + 9)^(1/2), x = -1 .. 1) );
```

The procedure **changevar** provides us with the facility to try several possible substitutions for a given integral, hoping that one of these may eventually work. If you find the syntax of the **changevar** command complicated, you may do it in two steps. For example, instead of the command

```
changevar( x^2 = u, Int(x * (9 - x^4) ^(1/2), x), u);
```

you may use the following two commands:

```
ex := Int(x * (9 - x^4) ^(1/2), x);
changevar( x^2 = u, ex, u);
```

Exercises with Maple

1. Consider the definite integral

$$\int_0^{1/2} \frac{x^2\,dx}{\sqrt{1-x^6}}\,.$$

a) See whether Maple can evaluate this directly.
b) Use the **evalf** command to obtain an approximation of the integral.
c) Try to find a substitution that works, and find the exact value of the integral.
d) Compare the answers to (b) and (c).

2. Consider the indefinite integral

$$\int \sqrt{1+\sin x}\;dx\,.$$

a) See whether Maple can evaluate this integral.
b) Try to find a substitution that works (think of trigonometric identities that might make the integrand a perfect square).
c) Check your answer by differentiating it to see whether you get the integrand back.

3* a) See whether Maple can evaluate $\int_0^{\pi} \frac{x \sin x}{1+\cos^2 x}\,dx$.

b) Use the substitution $u = \pi - x$ and appropriate trigonometric identities to express this integral in terms of $\int_0^{\pi} \frac{\sin x}{1+\cos^2 x}\,dx$. (Do this part by hand.)

c) Ask Maple to evaluate this last integral.

d) Substitute in the result of (b) to find the value of the original integral.

10 Inverses of Functions

In this lab session we will use Maple to find the inverse of a function f if it exists. Recall that the function g is called the inverse of f, if $f \circ g$ and $g \circ f$ are the identity functions; that is, if $(f(g(x)) = x$ and also $(g(f(x)) = x$. We can use Maple to determine whether two functions are inverses of each other. For example, if we define f, g, and h as

```
f := proc(x) x/(x - 3) end;

g := proc(x) 2*x - 5 end;

h := proc(x) 3*x/(x - 1) end;
```

then the following commands will show that f and g are not inverses of each other, whereas f and h are:

```
f(g(x))
simplify(");

g(f(x));
simplify(");

f(h(x));
simplify(");

h(f(x));
simplify(");
```

A function may or may not have an inverse. If f is a one-to-one function, then it does have an inverse, but we may or may not be able to find an explicit formula for the inverse function. One way to make sure that a function is one-to-one on a given interval is to verify that the sign of the first derivative remains positive or negative throughout the interval. The following steps may lead to the inverse of f:

Step 1. Solve the equation $y = f(x)$ for x. The Maple command is **solve(y = f(x), x);**. Note that for most functions, Maple may not be able to solve this equation. For some other functions, it may be able to find more than one solution (if the function f is not one-to-one). In yet other cases, it may produce real as well as complex solutions.

Step 2. If Maple is able to find a unique solution, we can find the inverse by replacing all the y's on the right-hand side by x and defining a function whose value at x is the expression on the right-hand side. (This can be done more conveniently using the **unapply** command in Maple.)

Example 1. Consider the function

$$f(x) = \frac{2x - 1}{x + 3} .$$

```
f := proc(x)  (2*x - 1)/ (x + 3) end;
```

We first check the derivative:

```
f1 := diff(f(x), x);
simplify(f1);
```

It is clear from this that $f'(x)$ is positive on $(-\infty, -3)$ and $(-3, \infty)$. This means that f is one-to-one throughout its domain, and hence the inverse must exist. To find it, we follow the foregoing instructions

```
solve( y = f(x), x);
subs ( y = x, ");
g := proc(x) - (3 * x + 1)/(x - 2) end;
```

In place of the last two commands we could have used the command **g := unapply(" , y)**. To verify that g is indeed the inverse of f, type

```
f(g(x));
simplify(");

g(f(x));
simplify(");
```

Since both compositions produced x, this confirms that g is in fact the inverse of f.

Example 2. Consider the function f defined by

$$f(x) = x^2 - 2x + 5.$$

```
f := proc(x) x^2 - 2 * x + 5 end;
```

Differentiate it.

```
f1 := diff(f(x), x);
```

Since the derivative is positive on $(1, \infty)$ and negative on $(-\infty, 1)$, the function f is not one-to-one on $(-\infty, \infty)$, but it would be if we restrict its domain to one of the foregoing intervals, say $(1, \infty)$. Now proceed as before:

```
solve(y = f(x) , x);
```

Since $x > 1$, we have to choose the solution $x = 1 + \sqrt{y-4}$:

```
g := proc(x) 1 + sqrt(x - 4) end;
```

Again, one can verify that g is the inverse by evaluating the compositions $f \circ g$ and $g \circ f$. (Do it.)

Example 3. Define f as

```
f := proc(x) sqrt((1 - x)/x) end;
```

The domain of this function is $(0, 1]$. As before, by looking at the derivative, we can conclude that f is increasing and hence one-to-one on this interval. So we continue with the foregoing steps:

```
solve( y = f(x), x);
g := proc(x) 1/(x^2 + 1) end;
```

Again, we can verify that $f \circ g$ and $g \circ f$ are both the identity functions. (Do it.)

Exercises with Maple

1. a) Verify that the function $f(x) = 2x - 7$ is one-to-one on its domain.
b) Find the inverse function g of f.
c) Verify that $f \circ g$ and $g \circ f$ are both identity functions, confirming that g is the correct function.

....

....

....

2. Repeat Exercise 1 for the function $f(x) = \dfrac{4x - 1}{x + 7}$.

....

....

....

3* Repeat for the function $f(x) = \sqrt{2 + 5x}$.

....

....

....

4. Repeat for the function $f(x) = \dfrac{\sqrt{4x^2 + 1} + 1}{2x}$.

....

....

....

11 Functions Defined by Integrals

In this lab we will study properties and graphs of functions that are defined as integrals. We know that if f is a continuous function on some open interval I, and a is a point in I, then the expression

$$F(x) = \int_a^x f(t)\, dt$$

defines a function F on I. Furthermore, by the Second Fundamental Theorem of Calculus, we have $F'(x) = f(x)$. Now if it is possible to find an antiderivative for f, then $F(x)$ can be expressed in terms of the known functions. But even if it is not possible to find an antiderivative for f, we can still investigate the function F and obtain valuable information about it. (Recall the definition and the properties of the natural logarithm function, for example.) Here is an another example:

Suppose the function F is defined by

$$F(x) = \int_1^x \frac{10 \sin t}{t+5}\, dt\,.$$

Find the absolute maximum and minimum of F on $[-4, 4]$, and sketch the graph of F on this interval. You will need pencil and paper to record the information, as you obtain it. It is not possible to find an antiderivative for the function $\frac{\sin t}{t+5}$ in terms of the known elementary functions. Since the function F is continuous on the closed interval $[-4, 4]$, it must have an absolute maximum and an absolute minimum on this interval. We know these will occur either at the critical points or the endpoints. To find the critical points, we calculate the derivative. By the Second Fundamental Theorem of Calculus, we have

$$F'(x) = \frac{10 \sin t}{t+5} \qquad [1]$$

This function is defined on the interval $[-4, 4]$, and it vanishes at $-\pi$, 0, and π in this interval. Hence these are the only critical points of F on $[-4, 4]$. To find the absolute maximum and minimum of F, we need only compare the values $F(-4)$, $F(-\pi)$, $F(0)$, $F(\pi)$, and $F(4)$. Since each of these is simply the value of a definite integral, we can use the built-in numerical integration capability of Maple to approximate these numbers. To organize things a bit, first we define F formally through the procedure

```
F := proc(x) int(10 * sin(t)/(t + 5) , t = 1 .. x) end;
```

Now we can find the approximate values of the five function values by

```
evalf(F(-4));
evalf(F(-Pi));
evalf(F(0));
evalf(F(Pi));
evalf(F(4));
```

A comparison of these values shows that F has the absolute maximum of (approximately) $5.3 = F(-\pi)$, and the absolute minimum of $-0.8 = F(0)$.

As for the graph of F on $[-4, 4]$, we see from [1], that

On $(-4, -\pi)$, $F'(x) > 0$ and hence F is increasing.
On $(-\pi, 0)$, $F'(x) < 0$ and hence F is decreasing.
On $(0, \pi)$, $F'(x) > 0$ and hence F is increasing.
On $(\pi, 4)$, $F'(x) < 0$ and hence F is decreasing.

It also follows that F has relative maxima at $x = -\pi$ and $x = \pi$, and a relative minimum at $x = 0$. We can have Maple evaluate F at a few other points. Then using the continuity and the foregoing information, we can sketch a pretty accurate graph of F.

```
evalf(F(-2));
evalf(F(-1));
evalf(F(1));
evalf(F(2));
```

Having done this with paper and pencil, we will now check to see whether Maple can draw the graph and whether it resembles ours:

```
plot(F(x), x = -4 .. 4);
```

Since the function evaluations require numerical integration, the execution of this command will be slow.

Exercises with Maple

1. The function F is defined by

$$F(x) = \int_0^x (t^2 - 4)e^{1-t^2}\,dt.$$

a) Find the absolute maximum and minimum of F on $[-3, 3]$.
b) Use the method discussed in this lab to sketch the graph of F on this interval.
c) Use Maple's **plot** command to compare your graph with Maple's. You may have to redraw portions of the graph on smaller intervals about the critical points.

....

....

....

12 Solving Differential Equations

An (ordinary) differential equation is a relation between the independent variable x, the dependent variably y, and various derivatives of y with respect to x. (In many physical applications, the independent variable is the time, denoted by t.) The rank of the highest derivative of y appearing in the equation is called the **order of the differential equation**. The following are some examples of differential equations:

$$y' + 3y = e^x$$

$$(x^2 + 1)y' = x^2 + y^2$$

$$y'' - 2y' + y = 3 \sin x.$$

The first two have order one, whereas the last equation has order 2. A solution to a differential equation is a function $y = f(x)$, so that when y and all its derivatives are substituted in the equation, the equation is satisfied. For example, the function

$$f(x) = \frac{1}{4}e^x + \frac{3}{4}e^{-3x}$$

is a solution of the first differential equation. Usually a differential equation has infinitely many solutions, but in many cases a unique solution may exist if appropriate initial conditions are given. For example, the function $f(x) = \frac{1}{4}e^x + \frac{3}{4}e^{-3x}$ is the unique solution to the first equation if the initial condition $f(0) = 1$ is required. The theory of differential equations is well developed, and many of you will probably take a course on the subject later. The purpose of this lab session is to see how the solution to some applied problems requires setting up differential equations, and how one can use Maple to get the solutions.

Example 1: Suppose an object with a temperature of 120 degrees is placed into a large freezer whose temperature of 25 degrees will remain the same throughout this experiment. Suppose that this object cools to 100 degrees after 5 minutes. What will the temperature of the object be after 10, 20, 30, and 60 minutes?

Solution: To set up a differential equation for this experiment, we will use a law from Physics. **Newton's Law of Cooling** states that the rate of change of a cooling body is proportional to the difference between the temperature of the body and that of the surrounding environment. So if we let $T(t)$ stand for the temperature of the object at time t, Newton's law of cooling gives

$$\frac{dT}{dt} = k\,(T - 25),$$

where k is the constant of proportionality. (This will be determined shortly, using the additional condition given in the problem.) The initial condition is

$$T(0) = 120.$$

Now that we have set up the differential equation and the initial condition,we can ask Maple to solve it:

```
dsolve({diff(T(t), t) = k * (T(t) - 25),T(0) = 120},T(t));
```

The command **dsolve** is for solving differential equations. The actual differential equation and the initial condition(s) are listed inside the braces. Note that the function name must be listed as $T(t)$, not just T. Now to find k, we use the additional information and substitute $t = 5$ and $T(t) = 100$ in the solution:

```
subs({t = 5, T(t) = 100}, ");
```

Then we solve this equation for k, and approximate it.

```
solve(", k);
evalf(");
```

So the function T is defined by

$$T(t) = 25 + 95\,e^{-0.4728t}.$$

Since we are going to find the value of $T(t)$ for several values of t, it may be more convenient to define it as a function:

```
T := proc(t) 25 + 95 * exp(-.04728 * t) end;
```

And now we have answers to all the questions asked in the problem:

```
T(10);
T(20);
T(30);
T(60);
```

An interesting additional question is: What is the limit of $T(t)$ as $t \to \infty$? Our intuition says (since the temperature of the freezer remains a constant 25 degrees) that after a long time, the temperature of the object should also be 25 degrees. Well, let's see:

```
limit (T(t), t = infinity);
```

Example 2: A tank contains 40 gallons of brine that includes 15 lb of salt. At the time $t = 0$ we add a brine containing 1.5 lb of salt per gallon to the tank, at the rate of 3 gallons per minute, and drain the well-mixed solution at the same rate. Find the amount of salt in the tank 30 minutes later.

Solution: Let $y = y(t)$ represent the amount of salt (in pounds) in the tank at time t. Then we have the initial condition $y(0) = 15$. The rate of change of the salt in the tank, $\frac{dy}{dt}$, is obtained from

$$\frac{dy}{dt} = r_1 - r_2,$$

where r_1 is the rate at which the salt enters the tank and r_2 is the rate at which the salt leaves the tank. Since in each minute 3 gallons of brine enter the tank, and each gallon contains 1.5 lb of salt, we get

$$r_1 = (3)\,(1.5) = 4.5 \text{ lb/min}.$$

On the other hand, at time t there are still 40 gallons of solution in the tank, containing $y(t)$ lb of salt, so each gallon contains $\frac{y(t)}{40}$ lb of salt, and since the solution is leaving the tank at the rate of 3 gal/min, we have

$$r_2 = 3\,\frac{y(t)}{40} \text{ lb/min}.$$

Hence we have

$$\frac{dy}{dt} = 4.5 - \frac{3y(t)}{40}.$$

We will use Maple to solve this differential equation subject to the initial condition $y(0) = 15$:

```
dsolve({diff(y(t), t) = 9/2 - 3*y(t)/40, y(0) = 15}, y(t));
```

Maple finds the solution $y(t)$ for us. It can be verified easily that this does in fact satisfy the differential equation and the initial condition. Now we can determine how many pounds of salt will be in the tank after 30 minutes.

```
subs(t = 30, ");
evalf(");
```

Both of the differential equations obtained in the foregoing examples were very simple. Maple can solve much more complicated differential equations, as well as systems of differential equations.

Exercises with Maple

1. Suppose we make the following changes in Example 1 of Lab 12:

a) The temperature of the freezer is 30 degrees and will remain that way.
b) The object cools to 85 degrees after 5 minutes.

Find $T(t)$, $T(20)$, $T(30)$, and $T(60)$.

....

....

2. Referring to Example 2 of Lab 12, assume that the well-mixed solution is being drained at the rate of only 2 gal/min. Set up the new differential equation and find y as a function of t. Be careful! In calculating the rate at which the salt is leaving the tank, the volume of the solution is no longer constant; it should depend on t. If the capacity of the tank is 100 gallons, what would the amount of salt be at the time of overflow?

....

3* Find the equation of a curve that passes through the point (0, 5) and has the property that the slope of the curve at every point P is twice the y-coordinate of P.

....

13 Integration by Parts, Reduction Formulas, and Pattern Recognition

One of the most frequently used techniques of integration is the method of **integration by parts**. This is the formula:

$$\int u\,dv = uv - \int v\,du.$$

By choosing u appropriately, one hopes that the application of this formula produces a simpler integral on the right-hand side. Maple's **student** package allows one to try several choices for u, to see whether one of them works. Once an appropriate choice for u is made, Maple will do all the routine work and produce the integral on the right-hand side. The name of the command is **intparts**, and it can be accessed using the **student** package:

```
with(student);
```

The command **intparts** has two arguments. The first argument is the desired integral in inert form (spelled with a capital *I*), and the second argument is the part that will be taken as u.

Example 1: To evaluate the integral

$$\int x \sin x\,dx,$$

we may first try choosing $u = \sin x$. To see the effect of applying the integration by parts formula, we type the following (make sure to spell "Int" with a capital *I*)

```
z := Int( x * sin(x), x);
intparts(z , sin(x));
```

We could have combined the two commands, but it is better this way since we can later refer to the same integral again by the name z. Now it appears from the answer to the last command that this choice of u was not good, since the integral on the right-hand side is more complicated than the original one. So we try another choice for u, $u = x$:

```
intparts(z , x);
```

The integral on the right-hand side is now trivial, and hence this *was* a good choice for u. Of course in general there is no guarantee that for a given integral some choice of u will work, but with Maple's help, we can afford to try many choices in a short amount of time. Sometimes integration by parts may have to be applied more than once in order to evaluate the integral successfully

Example 2: To evaluate the integral

$$\int e^x \sin x \, dx,$$

start by giving a name to this integral, and choosing $u = e^x$

```
z := Int( exp(x) * sin(x), x);
intparts(z , exp(x));
```

Now, the integral on the right-hand side is not necessarily simpler, but it is not any harder either. So we may try to apply integration by parts to that integral by choosing the new u as e^x again:

```
intparts(Int(exp(x) * cos(x),x), exp(x));
```

Now the integral on the right-hand side of this relation is exactly the original integral, which we called z. So we have

$$z = -e^x \cos x + e^x \sin x - z.$$

Solving this equation for z, we get

$$z = \frac{1}{2}(e^x \sin x - e^x \cos x).$$

We can ask Maple to differentiate this expression to see whether it is the right antiderivative

```
diff( 1/2 * ( exp(x) * sin(x) - exp(x) * cos(x)), x);
```

Example 3: Evaluate the integral

$$\int x^n e^x \, dx \, ,$$

where n is an arbitrary positive integer. We start by giving a name to the integral. Then we apply integration by parts by choosing $u = x^n$:

```
z := Int( x^n * exp(x) , x);
intparts(z , x^n);
```

It looks as if this can be simplified a little further:

```
simplify(");
```

Now the integral on the right-hand side is similar to the original integral, but the term x^n is changed to x^{n-1}. In other words, if we denote the original integral by $z(n)$, then we have shown that

$$z(n) = x^n e^x - n\, z(n-1). \quad [1]$$

This is an example of a reduction (or recursion) formula. One could try to apply integration by parts to the integral on the right-hand side, but this will not be necessary. One can simply replace all the occurrences of n in [1] by $n - 1$ and get

$$z(n-1) = x^{n-1} e^x - (n-1)\, z(n-2).$$

For a specific value of n, repeating this process n times will eventually produce the desired result. Formulas like [1] are useful for theoretical purposes, but in practice they are cumbersome. It is much more convenient to find an explicit formula for $z(n)$, if possible. Equipped with a computer algebra system (CAS) such as Maple, one would examine $z(n)$ for a few small values of z to see whether any particular pattern is emerging. We start by defining z as a function of n so that we can refer to values such as $z(2)$ or $z(n)$:

```
z := proc(n) int(x^n * exp(x), x) end;
```

Next, we will examine $z(n)$ for several small values of n. (We will also have Maple factor it for better appearance):

```
factor(z(1));
factor(z(2));
factor(z(3));
factor(z(4));
factor(z(5));
```

Now a close examination of these expressions reveals that

a) Each $z(n)$ is of the form e^x times a polynomial.

b) The polynomial part of $z(n)$ has degree n, and the terms of the polynomial are alternatingly positive and negative.

c) The first term of the polynomial part of $z(n)$ is x^n, and each succeeding term of that polynomial is simply the derivative of the preceding term. In other words, we are conjecturing that

$$z(n) = e^x[x^n - n\,x^{n-1} + n(n-1)x^{n-2} - \ldots + (-1)^n\, n!].$$

Is this a correct conclusion? Let's first verify this for some other value of n, say $n = 8$. According to the foregoing relation, we should have

$$z(8) = e^x(x^8 - 8x^7 + 56x^6 - 336x^5 + \ldots).$$

Let's see whether Maple confirms this

```
factor(z(8));
```

How about that! Of course, we have not proved anything yet; we have merely noticed that the foregoing explicit formula for $z(n)$ works for some selected values of n. It turns out that it is not very difficult to prove this. Induction, and the recursion formula [1], is all you need. Try it.

Exercises with Maple

1. Consider the integral

$$\int x^3 \sin x^2 \, dx.$$

Try using integration by parts for four different choices of u. List your choices here, and explain the problems with each choice, if any. Which choice led to the solution? Can you verify the solution by differentiating it?

....

....

....

2* Use integration by parts to evaluate $\int x^5 e^{x^2} dx$.

....

3. Assuming that n is a positive integer, consider the integral

$$\int x^n \ln x \, dx.$$

a) Have Maple evaluate this for small specific values of n, and then try to see whether you recognize a pattern. Can you guess the answer for general n?

....

b) Evaluate the integral, using integration by parts, with $u = \ln x$. Does the answer match your conjecture in (a)?

....

14 Integrals of Rational Functions

In the process of evaluating the integral of a rational function, there is a tedious step that involves writing the rational function as the sum of a polynomial and a collection of simple fractions. Each of these simple fractions has as its denominator a linear or an irreducible quadratic polynomial, or a power of such polynomials. This step is tedious in that it usually involves setting up a system of linear equations and then solving them. One of the forms of Maple's **convert** command can be used in this situation, provided Maple is able to factor the denominator as a product of linear and quadratic polynomials.

Example 1: Suppose we want to evaluate

$$\int \frac{x^2 - 3x + 1}{x^3 + x^2 - 2x}\, dx.$$

We should first give a name to the integrand:

```
ex := (x^2 - 3*x + 1)/(x^3 + x^2 - 2 * x);
```

The form of the **convert** command used for partial fractions requires the rational function as the first argument, the keyword "**parfrac**" as the second argument, and the variable x as the third argument:

```
convert(ex, parfrac, x);
```

It follows from this decomposition, that the original integral evaluates to

$$-\frac{1}{2}\ln x + \frac{11}{6}\ln(x+2) - \frac{1}{3}\ln(x-1) + C.$$

Example 2: Write the following rational function as a sum of partial fractions:

$$\frac{x^8 + 2x - 1}{(x-1)^3\,(x+3)^2}.$$

We ask Maple to convert it:

```
convert((x^8 + 2*x -1)/((x-1)^3 * (x + 3)^2), parfrac, x);
```

Maple quickly provides us with the result.

Example 3: Evaluate the integral

$$\int \frac{x^2\,dx}{x^4+1}.$$

```
convert(x^2/(x^4 + 1), parfrac, x);
```

Since Maple is unable to factor the denominator as a product of linear and quadratic polynomials with rational coefficients, it returns the original rational function. However, in this case we can factor the denominator as a product of two quadratic polynomials with real coefficients as follows:

$$\begin{aligned} x^4+1 &= x^4+2x^2+1-2x^2=(x^2+1)^2-\left(\sqrt{2}\,x\right)^2 \\ &= \left(x^2+\sqrt{2}\,x+1\right)\left(x^2-\sqrt{2}\,x+1\right). \end{aligned}$$

Now we can ask Maple to find the partial fractions:

```
convert(x^2/((x^2 + sqrt(2) * x + 1) *(x^2 - sqrt(2) * x + 1)),
parfrac,x);
```

Once again, Maple does the tedious calculations, and we have the desired factorization.

Example 4: Evaluate

$$\int \frac{x\,dx}{x^6+1}.$$

We try the **convert** command:

```
convert(x/(x^6 + 1), parfrac, x);
```

The first term is easy to integrate. As for the second, the denominator cannot be factored as a product of polynomials with rational coefficients, although it is possible to factor it as a product of polynomials with real coefficients, by completing the square. However, since the integrand has a special form, we will choose a different plan here. Go back to the original integral and make the substitution $u = x^2$:

```
with(student);
changevar(x^2 = u, Int( x /(x^6 + 1) , x));
```

Now convert this simpler rational function to partial fractions:

```
convert(1/(2*u^3 + 2), parfrac, u);
```

It is easy from here on, but to have the complete answer we ask Maple to integrate this last expression, and substitute in terms of x:

```
int(", u);
subs(u = x^2,");
```

Verify by differentiating:

```
diff(", x);
simplify(");
```

Exercises with Maple

1. a) Have Maple find the partial fraction decomposition of

$$\frac{x^3+1}{4x^6+12x^5+9x^4-4x^2-12x-9},$$

and use it to find the indefinite integral of this function.

b) Ask Maple to integrate the function directly, and compare the results.

....

2* Repeat Exercise 1 with the function

$$\frac{3x^4-2x^3+20x^2-5x+34}{(x-1)(x^2+4)^2}.$$

3. a) See whether Maple can find the partial fraction decomposition of

$$\frac{1}{x^4-x^2+4}.$$

b) Try to find a factorization of the denominator by completing the square. Then have Maple determine the partial fraction based on this factorization (see Example 3 of this lab). Finally, find an antiderivative for the first partial fraction.

15 Numerical Integration

In this lab we will consider the approximation of definite integrals by various methods. In Lab 8, we obtained two estimates for the value of

$$\int_a^b f(x)\,dx.$$

These estimates, known as the left and right sums, were merely the sums of the areas of rectangles. The width of each rectangle was $h = \dfrac{b-a}{n}$, and the height of the ith rectangle was the value of the function at the left or the right endpoint of the subinterval $[x_{i-1}, x_i]$. Since this process consisted of straightforward calculations, we created the Maple procedures **LeftSum** and **RightSum** to carry out these calculations. We noticed at that time that larger values for n resulted in better estimates for the value of the definite integral.

There are much better methods for estimating definite integrals. The **Trapezoidal rule** gives an approximation of the definite integral by adding up the areas of trapezoids built on each of the subintervals. **Simpson's rule**, on the other hand, approximates the integral by fitting parabolas through sets of three consecutive subdivision points and then adding up the areas under these parabolas. We have written two additional procedures that allow you to approximate the area using these two methods[1]. As with **LeftSum** and **RightSum**, these procedures, named **Trapezoid** and **Simpson**, require four arguments each: the function name, the lower limit, the upper limit, and the number of subdivisions. Note that in the case of Simpson's rule the number of subdivisions must be an even number. In this lab session we will examine the value of a definite integral using these four approximation methods. First we start by reading in these procedures:

```
read(`Sums.m`);
read(`Trapezoid.m`);
read(`Simpson.m`);
```

Example 1: Consider the integral

$$\int_1^3 \frac{dx}{(x+1)^2}\,.$$

We will approximate this using all four methods. First we define the function to be integrated:

f := proc(x) 1/(x + 1)^2 end;

`evalf(LeftSum(f, 1, 3, 20));`	This yields .2595571909
`evalf(RightSum(f, 1, 3, 20));`	This yields .2408071909
`Trapezoid(f, 1, 3, 20);`	This yields .250182190939121

[1] Later versions of Maple, including Maple V, have similar procedures included in the student package.

```
Simpson(f, 1, 3, 20);
```

This yields .250000400008024

Since we can actually find the exact value of the foregoing integral using the First Fundamental Theorem of Calculus, it would be interesting to see how these approximations differ.

Since an antiderivative for $\frac{1}{(x+1)^2}$ is $\frac{-1}{x+1}$, it turns out

that the exact value of the integral is $\frac{1}{4}$. A quick glance at the

foregoing four estimates shows that the Trapezoidal rule yields a better approximation than the left sum or right sum, and Simpson's rule yields better approximation than the Trapezoidal rule. This is usually the case. In the next lab we will examine the maximum error in using the Trapezoidal and Simpson's rules.

Example 2: Consider the integral

$$\int_0^2 e^{-x^2}\,dx.$$

```
f := proc(x) exp(-x^2) end;
```

```
evalf(LeftSum(f, 0, 2, 40));
evalf(RightSum(f, 0, 2, 40));
Trapezoid(f, 0, 2, 40);
Simpson(f, 0, 2, 40);
```

This yields .9066082434
This yields .8575240254
This yields .882066134089762
This yields .882081365321160

In this case it is not possible to find an antiderivative for the foregoing integrand in terms of the known functions, so we could not calculate the exact value. But we can ask Maple, using the **evalf** command, to get the value of the integral with ten digits of accuracy:

```
evalf(int(f(x), x = 0 .. 2));
```

This yields .8820813910, which again shows that Simpson's rule gave us the correct answer to about seven digits of accuracy.

Exercises with Maple

1. We know that

$$\int_0^1 \frac{4}{x^2+1}\,dx = 4\,[\tan^{-1}(1) - \tan^{-1}(0)] = 4\left(\frac{\pi}{4}\right) = \pi.$$

Use this to get four different approximations for the value of π (use $n = 40$). Then have Maple print the approximate value of π with 15 digits of accuracy and compare.

....

....

2. Approximate each of the following integrals by using the left sum, right sum, Trapezoidal rule, and Simpson's rule (choose $n = 16$ in all cases). Then compare your answer to the one obtained by Maple using the **evalf** command:

a) $\int_1^2 \frac{\sin x}{x}\,dx$ b) $\int_0^2 \sqrt{x^3+1}\,dx$

....

....

3* Use Simpson's rule with $n = 10$ to estimate the volume of the solid generated by revolving the region bounded by the curves $y = \sqrt[3]{1+x^3}$, $y = 0$, $x = 0$, and $x = 2$ about the x-axis.

....

....

16

Error Bounds in Trapezoidal and Simpson's Rules

We have seen that the Trapezoidal and Simpson's rules provide us with approximations to the value of a definite integral. In many cases we are interested in finding a bound on the error resulting from these approximations. It can be shown that in approximating

$$\int_a^b f(x)\,dx,$$

the maximum error committed is $\dfrac{(b-a)^3 T}{12\,n^2}$ for the Trapezoidal rule and $\dfrac{(b-a)^5 S}{180\,n^4}$ for Simpson's rule. Here T and S are the maxima of the absolute values of the second and the fourth derivatives of f on $[a, b]$, respectively.

Example 1: We will use the Trapezoidal and Simpson's rules (with $n = 20$) to approximate

$$\int_0^1 x^5\,dx.$$

```
read  (`Trapezoid.m`);
read  (`Simpson.m`);
f  := proc(x)  x^5  end;
Trapezoid(f,0,1,20);
```

This yields 0.16771 (when rounded to five decimal places).

The foregoing expression for the maximum error committed in using the Trapezoidal rule should now be defined as a function so that we can refer to it by name:

```
TrapMax  :=  proc(T,  a,  b,  n)    T  *  (b  -  a)^3/(12  *  n^2)  end;
```

To find the number T, we need to find the maximum of $f''(x) = 20x^3$ on [0, 1]. This is clearly 20, since $f''(x)$ is increasing on the interval [0, 1].

```
TrapMax(20,  0,  1,  20);
evalf(");
```

This shows that the maximum error is 0.00417 (rounded to five decimal places). So the exact value of the integral is in the range 0.16771 ± 0.00417, that is, between 0.16354 and 0.17188. Of course in this case we know the exact value of the integral. Since an antiderivative is $x^6/6$, the actual value is 1/6 or 0.16667, which certainly lies in this range.

Now we use Simpson's rule:

```
Simpson(f, 0, 1, 20);
```

This yields 0.16666875 (rounded to eight decimal places). Now we define the maximum error under the Simpson's rule as a function:

```
SimpMax := proc(S, a, b, n)  S * (b - a)^5/(180 * n^4) end;
```

To find the number S, we need to find the maximum of the absolute value of the fourth derivative of x^5 on [0, 1]. Since the fourth derivative is simply $120x$, we clearly have $S = 120$.

```
SimpMax(120, 0, 1, 20);
evalf(");
```

This yields 0.00000417 (rounded to eight decimal places). So the exact value of the integral must be in the range $0.16666875 \pm 0.00000417$, that is, between 0.1666646 and 0.1666728, which is certainly true.

Since Simpson's rule usually provides us with a better approximation, we will work only with that method in the rest of this lab session.

Example 2: Find the value of

$$\int_0^1 \sqrt{4 - x^2}\, dx,$$

with an error of at most 0.000001.

First we define the integrand:

```
f := proc(x) sqrt(4 - x^2) end;
```

We need to know what to take for n. The number S in the maximum error expression for Simpson's rule is the maximum of the absolute value of the fourth derivative of f on $[a, b]$. Let's have Maple calculate this derivative:

```
diff(f(x), x$4);
normal(");
```

We can find the maximum of the absolute value of this function on [0, 1] by finding its critical points and evaluating it at those and the endpoints. But let's have Maple plot the function instead, and read the maximum from the graph:

```
plot(", x = 0 ..1 );
```

It appears from the graph that the absolute value of the function attains its maximum at $x = 1$. Find out what it is, calculate the maximum error, and solve for n:

```
subs(x = 1, "");
S := abs(");
SimpMax(S, 0, 1, n);

fsolve (.000001 = " , n);
```

Since n has to be an even number, we take $n = 12$. In other words, we have shown that if we choose $n = 12$ with Simpson's rule, then the amount of error will be less than 0.000001. Let's verify this:

```
Simpson(f,0,1,12);
```

Now see whether Maple can evaluate the integral exactly:

```
int(f(x), x=0 .. 1);
evalf(");
```

The difference between these two numbers is clearly less than the desired number of 0.000001.

Example 3: Using Simpson's rule, find the value of

$$\int_1^2 \frac{\sin x}{x}\, dx$$

with an error of no more than 0.0000001.

```
f := proc(x) sin(x)/x end;
diff(f(x), x$4);
plot(" , x= 1 ..2);
```

It is clear from the plot that this function does not exceed $S = 0.15$ on the interval [1, 2].

```
SimpMax(.15, 1, 2, n);
fsolve (.0000001 = " , n);
```

So $n = 10$ should do it:

```
Simpson(f,1,2,10);
```

Since the foregoing integral cannot be evaluated exactly, we ask Maple to use its built-in numerical integration to find the value of the integral with ten digits of accuracy:

```
evalf(int(f(x), x = 1 .. 2));
```

The difference between these two numbers is certainly less than 0.0000001.

Exercises with Maple

1. a) What should the number n in Simpson's Rule be so that the error in approximating:

$$\int_0^5 \frac{x}{x^2+1}\,dx,$$

does not exceed 0.0001?

b) Find the exact value of the integral and compare it to the approximation by Simpson's rule using the value of n obtained in (a).

....

....

2 a) What should the number n in Simpson's rule be so that the error in approximating

$$\int_0^\pi \cos x^2\,dx$$

does not exceed .001?

....

b) Find the exact value of the integral to ten decimal places (using the **evalf** command) and compare it to the approximation by Simpson's rule using the value of n obtained in (a).

....

3* How large should n be to guarantee that the approximation of $\int_0^1 e^{x^2}\,dx$ by Simpson's rule is accurate to within 0.00001?

....

17 Graphs of Parametric Equations

Recall that the general form of the **plot** command provides for optional ranges for x and y. For example, to sketch the portion of the graph of $f(x)$ that lies in the rectangle $\{(x,y) \mid a \le x \le b, c \le y \le d\}$, the command `plot(f(x),x = a .. b, y = c .. d);` is used. In the absence of the ranges, Maple assumes $a = -10$, $b = 10$ and adjusts the values of c and d accordingly. There are many curves whose equations are either too complicated to put in the form $y = f(x)$ or do not define y as a single function of x. In many of these cases, the curve could be defined in parametric form. The parametric equations of a curve in the plane are usually given in the form:

$$x = f(t) \text{ and } y = g(t), \qquad [1]$$

where t varies over some interval. When t represents the time, equations in [1] give the position of a moving particle in the plane at time t. Maple's command for plotting graphs of parametric equations has the form `plot ([f(t) , g(t), t = a .. b]);` where optional ranges for x and y can be included just before the final parenthesis. The graph of equations in parametric form would look more natural if the units along the x- and y-axes are of equal size. Since the plot window of Maple is rectangular (with its width equal to about 60% of its length) to guarantee that the units along the all the rays will have the same size, we will always choose the ranges $a .. b$ and $c .. d$ in such a way that $(d - c)$ is about 60% of $(b - a)$.

Example 1: Sketch the graph of

$$x = 4 \cos t, y = 2 \sin t, 0 \le t \le 2\pi.$$

```
plot ([4*cos(t) , 2*sin(t), t = 0 .. 2*Pi], -5 .. 5, -3 .. 3);
```

As you may have expected, the curve is an ellipse whose standard equation in Cartesian coordinates can be obtained from the foregoing parametric equations by eliminating t.

Example 2: Sketch the graph of

$$x = \sec t, y = \tan t, -\pi/2 < t < \pi/2.$$

```
plot([sec(t), tan(t), t = -Pi/2 .. Pi/2], -5 .. 5, -3 .. 3);
```

Again, by eliminating t from the given parametric equations, one would obtain the equation $x^2 - y^2 = 1$ of a hyperbola. Why does the graph show only half this hyperbola?

Example 3: Sketch the graph of

$$x = 5\cos t + \cos 5t,\ y = 5\sin t - \sin 5t,\ 0 \le t \le 2\pi.$$

Let the range of values for x and y be $[-10, 10]$ and $[-6, 6]$, respectively.

```
plot([5*cos(t) + cos(5*t), 5*sin(t) - sin(5*t), t = 0..2*Pi],
-10..10, -6..6);
```

In this case it is not obvious at all that the parameter t can be eliminated to get the equation of the curve in Cartesian coordinates. However, this is a well-known curve called a six-cusped hypocycloid. For more information on hypocycloids and how they are generated, see Stewart.[1]

Exercises with Maple

1. For each of the following sets of parametric equations, first have Maple sketch its graph. Then convert each set to an equation in Cartesian form and see whether the graphs are as you expected

a) $x = 4\cos^2 t,\ y = 2\sin^2 t,\ 0 \le t \le \pi.$
b) $x = \cos^4 t,\ y = \sin^4 t,\ 0 \le t \le 2\pi.$
c)* $x = 4\sinh t,\ y = 3\cosh t,\ -3 \le t \le 3.$

....

....

....

2. Sketch the graph of the cycloid

$$x = t - \sin t,\ y = 1 - \cos t,\ 0 \le t \le 4\pi.$$

Choose the x-range nonnegative and large enough so that two full arches of the cycloid can be seen.

[1] James Stewart, *Calculus,* 2nd ed., Pacific Grove, CA: Brooks/Cole, 1991, p. 524.

18 Graphs in Polar Coordinates

Maple's **plot** command can be used to sketch the graph of functions in polar coordinates. Just as in the case of parametric equations, the graph of equations in polar coordinates would look more natural if the units along the x- and y-axes are of equal size. So as in the last lab we will choose the ranges a .. b and c .. d in such a way that $(d - c)$ is about 60% of $(b - a)$. The general form of the **plot** command for sketching the graph of $r = f(\theta)$, $\alpha \leq \theta \leq \beta$ is

```
plot([f(t),t,t = alpha..beta],a..b, c..d, coords = polar);
```

where `a..b`, and `c..d` are optional. We will now use this command to sketch several graphs in polar coordinates. (*Warning*: The plot command in polar mode can be quite slow!)

Example 1: Sketch the graph of

$$r = 3 \cos \theta + 2 \sin \theta. \qquad [1]$$

It is clear that we only need to choose θ between 0 and 2π to get the full graph

```
plot([3*cos(t) + 2*sin(t),t,t =0 .. 2*Pi], -5..5, -3..3, coords =
polar);
```

Note that the graph looks like a circle. In fact, by multiplying both sides of the equation [1] by r and converting to Cartesian coordinates, one can easily obtain a quadratic equation representing a circle.

Example 2: Sketch the graph of

$$r = 3 \sin \frac{\theta}{2}.$$

Applying the symmetry tests will show that the graph should be symmetric with respect to the x-axis, and the y-axis (and hence the pole as well). We may first try the command

```
plot([3*sin(t/2),t ,t = 0 .. 2*Pi],-5 .. 5, -3 .. 3,coords =
polar);
```

This graph does not appear to be symmetric about the y-axis! The problem lies in the fact that we have not allowed θ to vary in a large enough range to generate the whole graph. As θ is allowed to vary from 2π to 4π, $r = \sin \theta/2$ takes on negative values, generating another branch of the graph. Hence we need to choose the range for θ to be 0 to 4π, or equivalently, from -2π to 2π:

```
plot([3*sin(t/2),t,t = -2*Pi..2*Pi],-5..5, -3..3, coords = polar);
```

This one certainly has all three symmetries.

Example 3: Sketch the graphs of the following functions on the same set of axes:

$$r = 1 - 2\sin\theta$$
$$r = 1.$$

The first curve passes the symmetry test about the y-axis but not the x-axis. It is a well-known curve, called a Limaçon. The second curve is the unit circle with center at the pole.

```
plot({[1 - 2*sin(t), t, t = 0..2*Pi], [1, t, t = 0..2*Pi]},
-5..5, -3..3, coords = polar);
```

(Recall that if a command is too long to fit on a line, you may use the Return key to split it into several lines. To execute such a command, you need to select the whole command and then press the Enter key.)

Exercises with Maple

1* Have Maple sketch the graph of $r = \theta$ (the Spiral of Archimedes) for $0 \le \theta \le 6\pi$. Make sure to choose the ranges for x and y large enough to include all of the graph.

....

2. The curve $r = a \sin k\theta$ is a k-leaf rose if k is odd and a $2k$-leaf rose if k is even. Have Maple sketch the graph of $r = 3 \sin 4\theta$ on $[0, 2\pi]$. By dividing the domain into subintervals of length $\frac{\pi}{4}$, indicate the order in which the leaves are drawn as θ increases from 0 to 2π.

....

....

3. Have Maple sketch the graph of $r^2 = \cos 2\theta$ on $[0, 2\pi]$. Which symmetries did you expect the graph to have? What happens when θ is between $\frac{\pi}{4}$ and $\frac{3\pi}{4}$?

....

19 Rotation of Conic Sections

We know that the graph of the general quadratic equation

$$Ax^2 + Bxy + Cy^2 + Dx + Ey + F = 0 \qquad [1]$$

is always a conic section (circle, ellipse, parabola, or hyperbola). Degenerate cases include a point, a line, a pair of lines, or the empty set. When $B \neq 0$, a rotation of axes is required in order to put the equation in the standard form of the equation of a conic section. This means that the new equation in the variables X and Y will have no XY term. The procedure for obtaining the standard form is as follows:

If θ represents the angle of rotation, then we have $\cot 2\theta = \frac{A-C}{B}$. From this, we have to find the exact values of $\sin\theta$ and $\cos\theta$. Except in the very rare cases when this can be done mentally, one must use $\cot 2\theta$ to find $\cos 2\theta$, and then the trigonometric formulas

$$\sin\theta = \sqrt{\frac{1-\cos 2\theta}{2}} \quad \text{and} \quad \cos\theta = \sqrt{\frac{1+\cos 2\theta}{2}}$$

to find $\sin\theta$ and $\cos\theta$. Finally, the substitutions

$$x = X\cos\theta - Y\sin\theta \quad \text{and} \quad y = X\sin\theta + Y\cos\theta \qquad [2]$$

transform equation [1] to the desired form. As you may imagine, the calculations involved in this process are lengthy and error-prone. In this lab we will use Maple's computational capability to obtain the results faster.

Example 1. Transform the following quadratic equation so that the new equation has no XY term:

$$2x^2 + 24xy + 9y^2 - 8x - 6y + 10 = 0.$$

We start by defining the equation:

```
eqn := 2*x^2 + 24*x*y + 9*y^2 - 8*x - 6*y + 10 = 0;
```

Since we have $A = 2$, $B = 24$, and $C = 9$, the expression $\frac{A-C}{B}$ has a value of $\frac{-7}{24}$. We will need the arccot of this number. Unfortunately, for negative numbers Maple's built-in arccot function returns an angle in the fourth quadrant instead of the usual second quadrant. So knowing that cosine is negative in the second quadrant, we will include an extra minus sign in the command:

```
cos2t  :=  -  simplify( cos(arccot(-7/24)));
sint  := simplify(sqrt(1/2  -  1/2*cos2t));
cost  := simplify(sqrt(1/2  +  1/2*cos2t));
```

Note that **cos2t, sint**, and **cost** are just names and not function calls. Now we can substitute:

```
subs({x = cost*X - sint*Y, y = sint * X + cost * Y} , eqn);
```

Now to simplify, we use Maple's **radsimp** command, which works well with expressions involving radicals

```
radsimp(");
```

Maple produces the final form of the equation, and you can see that there is no *XY* term. If we wanted to put the equation in the standard form, we need to complete the squares at this point. This can be done using the function **completesquare**[1] of the **student** package.

```
with(student);
completesquare("", X);
completesquare(", Y);
```

In finding $\cos 2\theta$ from $\cot 2\theta$, and also in finding $\sin \theta$ and $\cos \theta$ from $\cos 2\theta$, one needs to calculate square roots. In Example 1, the coefficients of the quadratic equation were selected in such a way that in both cases the numbers were perfect squares. The next two examples show that some other cases can also be handled.

Example 2. Consider the quadratic equation

$$x^2 - 5xy + 13y^2 + 2x - 4y + 8 = 0.$$

```
eqn := x^2 - 5*x*y + 13*y^2 + 2*x - 4*y + 8 = 0;
```

Here $\frac{A-C}{B} = \frac{12}{5}$. The value of $\cos 2\theta$ will still be a perfect square, but the values of $\sin \theta$ and $\cos \theta$ will involve radicals. The same five commands, with the obvious change in the first, still work and produce the desired transformation:

[1] In previous releases (before Maple V), this command is **cmpltsq;**

```
cos2t  := simplify( cos(arccot(12/5)));
sint  := simplify(sqrt(1/2  -  1/2*cos2t));
cost  := simplify(sqrt(1/2  +  1/2*cos2t));
subs({x = cost*X  -  sint*Y,  y = sint*X  +  cost*Y},  eqn);
radsimp(");
```

Example 3. Consider the quadratic

$$x^2 + 3xy + 2y^2 - x + 2y + 5 = 0.$$

```
eqn := x^2 + 3*x*y + 2*y^2 - x + 2*y + 5 = 0;
```

Here $\frac{A-C}{B} = \frac{-1}{3}$. There will be radicals all over the place, but Maple still handles it reasonably well:[2]

```
cos2t  := -simplify(cos(arccot(-1/3)));
sint  := simplify(sqrt(1/2  -  1/2*cos2t));
cost  := simplify(sqrt(1/2  +  1/2*cos2t));
subs({x = cost*X  -  sint*Y,  y = sint*X  +  cost*Y},  eqn);
expand(");
```

Note that even though Maple shows the result with the *XY* terms, the two terms with *XY* in them actually cancel each other out! If you want to get an idea of what the coefficients are like, type

```
evalf(");
```

The nonzero coefficient for *XY* is the result of round-off error.

[2] If you are using a release earlier than Maple V, replace the **expand(");** command with **radsimp(");**

Exercises with Maple

1. Transform each of the following equations so that the new equation has no XY term:

a) $x^2 - xy + y^2 - 80y - 24 = 0.$

b) $25x^2 - 7xy + y^2 - 2x - 10y - 17 = 0.$

c)* $16x^2 - 8\sqrt{2}\,xy + 2y^2 + \left(8\sqrt{2} - 3\right)x - \left(6\sqrt{2} + 4\right)y = 7.$

....

....

....

20 Sums and Infinite Series

In this lab we work primarily with Maple's **sum** function to determine the exact values of finite sums as well as some convergent infinite series. This function can be used to print the value of a finite definite sum such as

$$\sum_{k=1}^{10000} (k^2 - 3k + 2)$$

```
sum(k^2 - 3*k + 2, k = 1 .. 10000);
```

If the upper limit of the summation is a variable, then we have an indefinite sum (in analogy with definite and indefinite integrals). In some cases, Maple can come up with an explicit formula for the indefinite sum:

```
sum(k, k = 1 .. n);
simplify(");
```

This is the well-known formula

$$1 + 2 + \ldots + n = \frac{n(n+1)}{2},$$

giving the sum of the first n positive integers. In fact, Maple can quickly find a formula for

$$\sum_{k=1}^{n} k^r$$

for small integer values of r. To see several examples in an efficient way, we define

```
power := proc(r) sum(k^r, k = 1 .. n) end;
```

Now we evaluate it for several values of r:

```
power(4);
power(10);
power(-1);
```

The last command gives the answer in terms of the built-in function **psi**, and the constant gamma. For a specific value of n, one can find the approximate value of the sum (with ten digits of accuracy):

```
evalf(subs(n = 200, "));
```

This is the sum of the first 200 terms of the Harmonic series. If the name "infinity" is used as the terminal value of k, we have an infinite series:

```
sum(1/k^(1/2), k = 1 .. infinity);
```

Since this series is divergent (by the p-series test), Maple just prints the series in sigma-notation.[1] On the other hand, the command

```
sum(1/k^2, k = 1 .. infinity);
```

gives the exact value of this convergent series, whereas the command

```
sum(1/k^3, k = 1 .. infinity);
```

gives the sum of this convergent series in terms of the built-in **Zeta** function which can be approximated using

```
evalf(");
```

If Maple cannot find the exact value of a convergent series or cannot express it in terms of built-in functions such as **psi** or **Zeta**, it prints the series in sigma-notation:

```
sum(1/k^(3/2), k = 1 .. infinity);
```

One can always find the nth partial sum and evaluate it at a desired n:

```
PS := proc(n)  sum(1/k^(3/2), k = 1 .. n) end;
evalf(PS(200));
```

Now consider a convergent alternating series such as

```
sum( (-1)^k/k^2, k = 1 .. infinity);
```

Maple prints the answer in sigma-notation again.[2] But using the partial sums, not only can we approximate the sum but we can also determine the truncation error.

---

[1] Maple V gives the answer Zeta(1/2) to this command. But if you follow it with the command `evalf(");` it gives a negative decimal number, which we know to be incorrect. The series is certainly divergent.

[2] Maple V tries harder, and gives the answer in terms of the hypergeometric function. Follow this command with `evalf(");` to see what happens.

```
PS := proc(n)   sum((-1)^k/k^2), k = 1 .. n) end;
evalf(PS(100));
```

Since the truncation error in a convergent alternating series is less than the absolute value of the next term, we conclude that this answer is correct with an error of less than 0.0001.

Exercises with Maple

1. a) Have Maple find the value of the infinite series

$$\sum_{k=1}^{\infty} \frac{1}{k^2 + 3k + 2}.$$

b) See whether you can get the same answer mathematically by writing the kth term as a sum of partial fractions and then forming a telescopic sum for the sum of the first n terms.

....

....

2. Maple can detect whether a series is a geometric series and can find the sum immediately if the series is convergent. Try the following series to see whether Maple gives the expected answers:

$$\sum_{k=1}^{\infty} \left(\frac{-4}{3}\right)^k \qquad \sum_{k=1}^{\infty} \left(\frac{2}{3}\right)^k \qquad \sum_{k=1}^{\infty} (-1)^k \qquad {}^*\sum_{k=0}^{\infty} 5\left(\frac{e}{3}\right)^k$$

(You can use E or exp(1) for e in the last sum.)

....

3. The following series satisfies the Alternating Series Test and is therefore convergent. Find its sum with an error of less than 0.00001.

$$\sum_{k=1}^{\infty} \frac{(-1)^k}{k^{5/2}}$$

....

21 Taylor Polynomials and Function Evaluation

Among all mathematical functions, polynomial functions are the easiest to evaluate. Thus it is desirable to approximate the value of a function f at a point a by evaluating an appropriate polynomial at a. For a nonnegative integer n, the nth-degree Taylor polynomial of f at a is a polynomial $P_n(x)$ with the property

$$P_n^{(k)}(a) = f_n^{(k)}(a), \quad \text{for } k = 0, 1, 2, \ldots, n.$$

These $n + 1$ conditions determine the polynomial P_n uniquely as

$$P_n(x) = \sum_{k=0}^{n} \frac{f^{(k)}(a)}{k!}(x-a)^k.$$

Maple's **taylor** command has the form **taylor(f(x), x = a, n);** but actually produces the Taylor polynomial of degree $n - 1$. It also appends an error term of the form O(n) to the end. In order to produce the Taylor polynomial alone, we must use a form of the **convert** command. Let's start by defining a function f:

```
f := proc(x) sin(x) + cos(x) end;
```

To minimize typing, we will define the nth-degree Taylor polynomial as a function of n. For simplicity, we choose $a = 0$:

```
TayPoly:= proc(n) convert( taylor(f(x), x = 0, n+1), polynom) end;
```

Now the following three commands produce the second-, fifth-, and thirteenth-degree Maclaurin polynomials of f, respectively:

```
P2  := TayPoly(2);
P5  := TayPoly(5);
P13 := TayPoly(13);
```

Let us use these polynomials to approximate the value of

$$f\left(\frac{\pi}{4}\right) = \sin\left(\frac{\pi}{4}\right) + \cos\left(\frac{\pi}{4}\right) = \sqrt{2} \approx 1.414213562.$$

Substituting $x = \frac{\pi}{4}$ into each of the foregoing polynomials,

we get

```
evalf (subs ( x = Pi/4 , P2));
evalf (subs ( x = Pi/4 , P5));
evalf (subs ( x = Pi/4 , P13));
```

We observe that polynomials of higher degree give rise to better approximations of the value of the function. It is also informative to sketch the graph of f and each of the polynomials on the same coordinate axes and observe how the graphs of these polynomials approach the graph of f in an appropriate interval about $x = a$:

```
plot({f(x), P2}, x = -Pi .. Pi);
plot({f(x), P5}, x = -Pi .. Pi);
plot({f(x), P13}, x = -2*Pi .. 2*Pi);
```

As another example, we will ask Maple to produce the Taylor polynomials of various degrees for the function $f(x) = e^{x - \pi/4} \tan x$:

```
f := proc(x) exp(x - Pi/4) * tan(x) end;
P4 := TayPoly(4);
P16 := TayPoly(16);
P50 := TayPoly(50);
```

To produce this last polynomial by hand would have been extremely tedious, to say the least! Now let's get an approximate value of $f(\frac{\pi}{4}) = 1$:

```
evalf(subs(x = Pi/4, "));
```

How much error do we commit when we approximate f by one of these polynomials? According to the Taylor Remainder Theorem, the error is given by

$$R_n(x) = \frac{f^{(n+1)}(c)}{(n+1)!}(x-a)^{n+1},$$

where c is a number between a and x. For example, when we used P_5 to approximate $f\left(\frac{\pi}{4}\right)$ for the function $f(x) = \sin x + \cos x$, the error was

$$R_5\left(\frac{\pi}{4}\right) = \frac{f^{(6)}(c)}{6!}\left(\frac{\pi}{4}\right)^6.$$

Now since all the derivatives of f are of the form $\pm \cos x \pm \sin x$, it follows that regardless of the value of c, the absolute value of $f^{(6)}(c)$ cannot exceed 2, and hence we have

$$\left|R_5\left(\frac{\pi}{4}\right)\right| \leq \left(\frac{2}{6!}\right)\left(\frac{\pi}{4}\right)^6 \approx 0.0006511.$$

The actual error was about .0003586.

Using the reverse of this process, given a maximum tolerance level, one can determine how large the number n would have to be in order for the error not to exceed that tolerance level (see Exercise 2 of this lab).

The rate of convergence of the Taylor series should also be an important consideration. For example, it can be shown that the interval of convergence for the function $f(x) = 4 \arctan x$ is $[-1, 1]$. However, to use the Taylor polynomials (at $x = 0$) to approximate the value of this function at 1 (which is the number π) is not very efficient:

```
convert(taylor(4 * arctan(x), x = 0, 100), polynom);
evalf(subs(x = 1 , "));
```

Even if we increase n to 500 or even to 1000, the approximation obtained for π will still be quite poor. (Try it!)

Exercises with Maple

1* Find the Taylor polynomials of degree 3, 10, and 20 for the function $f(x) = e^x \cos 2x$ about the point $x = 0$.

....

2. For the function $f(x) = \sin x + \cos x$ discussed in the lab, how large should n be to guarantee that the error in calculating $f\left(\frac{\pi}{4}\right)$ will not exceed .0000001?

....

3. a) Find the Taylor polynomials of degrees 5, 8, and 25 for $f(x) = \tan x$ about $x = 0$.

b) Use these polynomials to get three different approximations for tan(35°).

c) Can the polynomials in (a) be used to approximate $\tan \pi$? (Why?) Try it!

d) Find the maximum error in using the 8th-degree Taylor polynomial to approximate tan(35°). (*Hint*: Use the graph of the 9th derivative to find its maximum on the interval under consideration.)

e) Find the actual error by having Maple use its internal evaluation of tan(35°) to ten significant digits, and verify that the actual error is in fact less than the maximum error.

....

....

....

....

....

22 Vectors in Maple

In this lab we will learn some of the useful Maple commands that can be used to carry out calculations with vectors. In Maple, vectors are represented as one-dimensional arrays. For example, the command

```
v := array ([2, -3, 7]);
```

defines **v** as a vector in 3-space with components 2, –3, and 7. Note that in echoing this assignment, Maple also confirms the fact that **v** is a vector in 3-space by indicating that **v** is of the type array(1..3). In fact, Maple treats vectors as special cases of matrices, which are 2-dimensional arrays of numbers. For example, the command

```
A := array ([ [1, 2, 3] , [4, 5, 6] ]);
```

defines **A** as a matrix with two rows and three columns whose rows are the vectors listed within the outer brackets. Maple can manipulate vectors and matrices by performing useful operations on them. Most of these operations are defined in a package called **linalg**. As usual, we load this package using a **with** statement:

```
with (linalg);
```

As you can see from the list of procedures printed out on the screen, there are numerous such functions. We will use only a few of these procedures in this lab session.

Addition of vectors: If we define another vector **u** by

```
u := array ([2 , 3, 5]);
```

then the "sum" of these vectors is assigned to **w** by

```
w := add(u, v);
```

A more general version of this command allows one to form a linear combination of the two vectors with given coefficients. For example, the command

```
add(u, v, 2, 3);
```

determines the vector 2**u** + 3**v**. In particular, to find the difference **u** – **v**, we can type

```
add(u, v, 1, -1);
```

The scalar product of two vectors: Also known as the *inner product*, or the *dot product* of two vectors. The scalar product can be obtained by the command

```
dotprod(u , v);
```

The length (norm) of a vector: The norm of the vector **v** is just the square root of the dot product of **v** with itself. However, the command

```
norm(v , 2);
```

can be used to calculate it directly.

The cross product of two vectors: Also known as the external product of **u** and **v**, this is a vector perpendicular to both **u** and **v**, and has a magnitude numerically equal to the area of a parallelogram having **u** and **v** as its adjacent sides. The cross product is usually denoted by $\mathbf{u} \times \mathbf{v}$. It is defined only for vectors in 3-space, and can be determined using the command

```
crossprod(u, v);
```

The angle between two vectors: The angle θ between **u** and **v** satisfies the relation

$$\mathbf{u} \cdot \mathbf{v} = \|\mathbf{u}\| \, \|\mathbf{v}\| \cos \theta,$$

and can be calculated using the command

```
angle(u , v);
```

To get an approximate value for this angle, type

```
evalf( " * 180/Pi);
```

Example 1: Find the area of the triangle whose vertices are $A(1, 2, 4)$, $B(3, 2, 5)$, and $C(2, 2, 0)$.

The vectors $\mathbf{u} = \overrightarrow{AB} = (2, 0, 1)$ and $\mathbf{v} = \overrightarrow{BC} = (-1, 0, -5)$ form the two adjacent sides of a parallelogram whose area is twice the area of the triangle ABC. On the other hand, the area of this parallelogram is precisely the magnitude of the cross product of **u** and **v**. Hence we proceed as follows:

```
u := array([2, 0 , 1]);
v := array([-1, 0 , -5]);
w := crossprod(u, v);
area := 1/2 * norm(w, 2);
```

Example 2: Find the equation of the plane passing through the three points A, B, and C in Example 1.

Recall that the vector **w** determined in Example 1 is perpendicular to the plane through A, B, and C. Hence the point $P(x, y, z)$ is on this plane if and only if the vector $\overrightarrow{AP}$ is perpendicular to **w**. Since we know the components of $\overrightarrow{AP}$, we define it:

```
AP := array ([x - 1, y - 2, z - 4]);
```

And now we can find the equation of the plane by forcing $\overrightarrow{AP}$ to be perpendicular to **w**:

```
eqn := dotprod(AP, w) = 0;
```

When simplified, this gives the equation $y = 2$, which is the correct plane by direct observation.

Alternative Solution to Example 2

It can be shown that if the points: $P_i = (x_i, y_i, z_i)$, $i = 1, 2, 3$, are not on the same line, then the equation of the plane through these points is given by

$$\begin{vmatrix} x & y & z & 1 \\ x_1 & y_1 & z_1 & 1 \\ x_2 & y_2 & z_2 & 1 \\ x_3 & y_3 & z_3 & 1 \end{vmatrix} = 0.$$

So, we define the corresponding 4 by 4 matrix and have Maple calculate its determinant:

```
B:= array([[x, y, z, 1], [1, 2, 4, 1], [3, 2, 5, 1], [2, 2, 0,
1]]);
eqn := det(B) = 0;
```

This gives us the same equation as before.

Example 3: Another useful command in the **linalg** package is **linsolve(A,b)**, which determines the solution(s) to a system of m linear equations in n unknowns. Here **A** is the matrix of coefficients, and **b** is the vector whose components are the numbers on the right-hand side of the equations. The only requirement is that the number of rows of **A** must match the number of components of **b**, which has to be the case anyway. Of course, depending on m and n and the actual equations, there may be none, one, or infinitely many solutions. Let's solve the following system of four equations in four unknowns:

$$\begin{cases} 3x - 2y + 3z - t = 6 \\ 2x - 17y + \quad 3t = 34 \\ 5x + 2y - z - 2t = -9 \\ \quad 3y + z + t = 4 \end{cases}$$

```
A := array([[ 3, -2, 3, -1], [2, -17, 0, 3], [5, 2, -1, -2], [0,
3, 1, 1]]);
b := array( [6, 34, -9, 4]);
linsolve(A,b);
```

Maple gives the solution almost immediately. If a system has infinitely many solutions, Maple gives the complete solutions in terms of one or more arbitrary parameters.

Exercises with Maple

1. a) Verify that the triangle with vertices at $A(0, -2, 7)$, $B(2, -1, 3)$, and $C(-1, 1, 2)$ is a right triangle. Is this triangle isosceles? (Why?)

b) The volume of the parallelepiped whose three adjacent sides are vectors **u**, **v**, and **w** is given by **u** . (**v** × **w**). Find the volume of the parallelepiped whose adjacent sides are three vectors $\overrightarrow{OA}$, $\overrightarrow{OB}$, and $\overrightarrow{OC}$. (A, B, and C as above). How would you check whether three given vectors are coplanar (that is, lie in the same plane)?

....

....

....

2. Find the equation of the plane passing through the points $A(3, 4, 0)$, $B(0, 2, 2)$, and $C(2, 0, 1)$.

....

3* Find the equation of the plane that passes through $(-1, 0, 1)$ and contains the line $x = 5t$, $y = 1 + t$, $z = -t$.

....

23 Plotting Graphs in Three-dimensional Space

The graph of the function $f(x, y)$ of two variables is defined to be the set of all points (x, y, z) in 3-space with the property that $z = f(x, y)$. This graph is called a **surface**. Except in some very simple cases, sketching the graph of a surface by hand is quite tedious and may not resemble anything like the true shape. Frequently one visualizes the shape of a surface by considering its traces or cross sections with planes parallel to the coordinate planes. Maple V's 3-dimensional plotting routines facilitate the drawing of surfaces in Cartesian, cylindrical, and spherical coordinate systems. In Cartesian coordinates, we can draw surfaces whose equations are in the form $z = f(x, y)$, as well as those whose equations are given in the parametric form:

$$x = f(r, s)$$
$$y = g(r, s)$$
$$z = h(r, s).$$

Also note that in Cartesian coordinates, an equation in which one of the symbols x, y, or z is missing represents a right cylinder with generator parallel to that axis.

Example 1: Sketch the graph of

$$z = \sin x + \cos 2y, \text{ over the square: } -\pi \le x \le \pi, -\pi \le y \le \pi:$$

```
plot3d(sin(x) + cos(2*y), x = -Pi .. Pi, y= -Pi .. Pi);
```

After viewing the graph, you may redraw it by changing some of the plot options. To do this, choose the 3D Plot Options menu item from the Format menu and modify Axis (None, Box, Frame, Normal), Plot Style (Hidden, Wire, Patch), Color Scheme, View Angles, and other characteristics.

Example 2: Sketch the graph of

$$z = \sin\left(\sqrt{x^2 + y^2}\right).$$

Since the equation is already solved for z, we need only choose the x and y ranges. Let's draw it over the square: $6 \le x \le 6$ and $6 \le y \le 6$.

```
plot3d(sin(sqrt(x^2 + y^2)) , x = -6.. 6, y= -6 .. 6 , style =
PATCH);
```

Example 3: Sketch the graph of the following surfaces over an appropriate domain:

a) $y = 5 \sin \frac{x}{2} \cos \frac{2x}{3}$

b) $z = \frac{2 - y}{y^2 + 1}$.

The equation in (a) is missing the variable z, so its graph is a cylinder with generator parallel to the z-axis. Furthermore, this can be viewed as the parametric equation of a surface, with x and z as parameters; that is,

$$x = x,\ y = 5 \sin \frac{x}{2} \cos \frac{2x}{3},\ z = z$$

```
plot3d([x, 5*sin(x/2)*cos(2*x/3), z], x = -10 .. 10, z = -10 ..
10, style = PATCH);
```

Since the variable x is missing in equation (b), the graph must be a cylinder with generator parallel to the x-axis. Since the equation is solved for z, we can use the first version of the **plot3d** command:

```
plot3d((2 - y)/(y^2 + 1), x = -10 .. 10, y = -10 ..10);
```

Example 4: Sketch the graph of the ellipsoid

$$\frac{x^2}{4} + \frac{y^2}{9} + z^2 = 1.$$

This equation does not define z as a function of x and y. However, by solving the equation for z, we get two separate functions, the graph of each function being half of the ellipsoid. We will have Maple do the required calculations for plotting the two graphs, and then use the **display3d** command to see them simultaneously:

```
G1 := plot3d( sqrt(1 - x^2/4 - y^2/9), x = -2 .. 2 , y = -3 .. 3):
G2 := plot3d(-sqrt(1 - x^2/4 - y^2/9), x = -2 .. 2 , y = -3 .. 3):
```

Note that the foregoing commands end with a colon. Now to use the **display3d** command, we need to load the **plots** package. This package also includes the **cylinderplot** and

sphereplot commands, to be discussed shortly.

```
with(plots);
```

Now we are ready to display the two halves of the ellipsoid simultaneously:

```
display3d({G1 , G2});
```

We will next see how Maple can draw graphs of surfaces in cylindrical coordinates. If the equation of the surface is of the form $r = f(\theta, z)$, the command `cylinderplot(f(θ, z), θ = a..b, z = c..d);` will do. However, Maple also allows for the parametric form of the equation, in which case it is assumed that r, θ, and z are expressed in terms of two arbitrary parameters u and v. In that case, the command `cylinderplot([r(u,v), θ(u,v), z(u,v)], u = a..b,v = c..d)` should be used.

Example 5: Sketch the graph of each of the following surfaces in cylindrical coordinates:

a) The cylinder $r = 3$

b) The half-plane $\theta = \dfrac{3\pi}{4}$

The first equation is solved for r, so we use

```
cylinderplot(3 , t = 0 .. 2*Pi , z = -5 .. 5, style = PATCH);
```

The second equation can be converted to the parametric form, where r and z are selected as parameters:

```
cylinderplot([r, 3*Pi/4, z], r = 0..5, z = -4 .. 4, axes = BOXED);
```

The commands for plotting in spherical coordinates are similar. If the equation of the surface is of the form $\rho = f(\theta, \varphi)$, the command `sphereplot (f(θ, φ), θ = a..b, φ = c..d);` will do. However, Maple also allows for the parametric form of the equation, in which case it is assumed that ρ, θ, and φ are expressed in terms of two arbitrary parameters u and v. In that case, the command `sphereplot ([ρ(u,v), θ(u,v), φ(u,v)], u = a..b, v = c..d);` should be used.

Example 6: Sketch the graph of each of the following surfaces in spherical coordinates:

a) The sphere $\rho = 5$

b) The cone $\varphi = \dfrac{\pi}{6}$

Based on the foregoing comments, the following commands should produce the corresponding graphs:

```
sphereplot(5 , t = 0..2*Pi, phi = 0..Pi, style = PATCH, axes =
BOXED);
sphereplot([r , t, Pi/6], r = 0..4, t = 0 .. 2*Pi);
```

Both of these graphs will look more realistic if the **Constrain View** box (in **3D plot Options**) is checked. The cone looks better if the **Wire** style is used.

Exercises with Maple

After drawing the following surfaces, your instructor may ask you to save each image as a PICT to an appropriately named file, and turn in a disk containing these files before you leave the Lab.

1* Have Maple sketch the graph of $f(x, y) = 1 - x^2$ over an appropriate domain.

2. Draw the graph of

$$z = \frac{\sin(x^2 + y^2)}{x^2 + y^2 + 1}.$$

Select the style and the axis options in such a way that the graph looks the best. Does the graph look better with the Constrain View box checked?

3. Draw the graph of $r = 5 \cos 2\theta$ in cylindrical coordinates.

4. Draw the graph of the surface $r = \sin\theta + \cos\varphi$ in spherical coordinates. Which choices of the plotting options give rise to the best graph?

....

5. a) Draw the graphs of the sphere with radius 5 and the cone $\varphi = \pi/6$ on the same axes. Make sure that the visible part of the cone extends beyond the sphere.

b) Enhance the graph in (a) to include an inverted cone (of the same size) extending out of the lower section of the sphere.

24 Functions of Several Variables (Part A)

The main concepts of calculus–such as limit, differentiation, and integration–can be extended in a natural way to functions with several variables. In this lab session we will practice with the Maple commands that facilitate the manipulation of such functions. We will primarily work with functions with two independent variables, but functions with three or more variables can be handled in a similar way. A function of two variables is declared basically the same way as a single-variable function. For example, to define the function

$$f(x, y) = \frac{2xy}{x^2 + y^2},$$

we use the command

```
f := proc(x,y) 2*x*y/(x^2 + y^2) end;
```

Now we can easily evaluate this function at any point:

```
f(1,2);
f(x,y);
f(0,0);
```

The **limit** command can be used to find the limit of a function of several variables, if it exists. For example, to find the limit of the foregoing function at (1, 2), we use

```
limit(f(x,y), { x = 1 , y = 2});
```

Since f is continuous at every point other than the origin, its limit at any of these points is simply the value of the function at that point. Let us now try the limit at (0, 0):

```
limit(f(x,y), { x = 0, y = 0});
```

Maple tells us that the limit does not exist. This can be verified directly by finding the limit along the x-axis ($y = 0$) and along the line $y = x$, and observing that two different values are obtained. In fact, by switching to polar coordinates one can see that any value between -1 and 1 can be obtained. (Try it.)

The **limit** command does not always give the correct answer to the limit. For example, consider the function

$$g(x, y) = \frac{x^3 y}{x^6 + y^2}.$$

```
g := proc(x, y) x^3*y/(x^6 + y^2) end;
```

This function is also continuous at every point except at (0, 0). Let us see what happens if we try to find the limit at this point:

```
limit (g(x,y), {x = 0, y = 0});
```

Maple V echos the command. Earlier releases of Maple might tell you that the limit is zero. But is it really? The limit along the x-axis or along any straight line through the origin is clearly zero. However, the limit along the curve $y = x^3$ is actually $\frac{1}{2}$, implying that the limit at (0, 0) cannot possibly exist.

Maple's **diff** command can be used to find the partial derivatives of a function of several variables. For example,

```
diff ( g(x,y) , x);
normal(");
```

produces the simplified expression for $\frac{\partial g}{\partial x}$. For a higher partial derivative such as

$$\frac{\partial^5 g}{\partial^2 x\, \partial^3 y},$$

we can use the command

```
diff(g(x,y), x$2, y$3);
```

Maple can also be used to find the directional derivative of a function of several variables along a given vector. The **directional derivative** of the function $f(x, y)$ along the vector **v** is simply the dot product of the gradient of f and the unit vector along **v**. The gradient (which is a vector with components $\frac{\partial f}{\partial x}$ and $\frac{\partial f}{\partial y}$) can be calculated using the linear algebra package.

```
with(linalg):
```

grad, the name of the function in this package that returns the gradient of a function, can be invoked using the format

```
Gr := grad(f(x,y), [x , y]);
```

Remember that *f* refers to the first function we defined in this lab session. To find the directional derivative of *f* along the vector **v** = 3 **i** – 4 **j**, first we define the vector **v**:

```
v := array([3 , -4]);
```

Now we simply ask Maple to form the dot product of **Gr** and the unit vector along **v** and simplify it:

```
dotprod (Gr , v) / norm(v , 2);
normal(");
```

Recall that norm(v, 2) is just the length of the vector **v**. To find the directional derivative of *f* at (1, 2) along the vector **v**, we can now use the command

```
subs(x = 1, y = 2 , ");
```

Exercises with Maple

1*. Let $f(x, y) = \dfrac{2x^2y}{x^4 + y^2}$.

a) Find the limit of $f(x, y)$ at $(2, -1)$.
b) Does the limit at $(0, 0)$ exist? (Why?)

c) Find $\dfrac{\partial^7 f}{\partial^5 x\, \partial^2 y}$ at $(1, 0)$.

d) Find the directional derivative of f at $(0, 2)$ along the vector $\mathbf{v} = 5\,\mathbf{i} + 12\,\mathbf{j}$.

....

....

....

....

2. Let $f(x, y, z) = \dfrac{5xy}{x^2 + 2y^2 + 3z^2}$.

a) Find the limit of $f(x, y, z)$ at $(1, 2, -1)$.
b) Does the limit at $(0, 0, 0)$ exist? (*Hint*: First find the limit along the coordinate axes, then along another line through the origin.)

c) Find $\dfrac{\partial^9 f}{\partial^2 x\, \partial^3 y\, \partial^4 z}$ at $(1, 2, 3)$.

d) Find the directional derivative of f at $(0, 2, 1)$ along the vector $\mathbf{v} = 2\,\mathbf{i} + \mathbf{j} - 2\,\mathbf{k}$.

....

....

....

25 Functions of Several Variables (Part B)

In this lab we will continue working with functions of several variables. In particular, we will learn how to find the relative extrema and saddle points of a function of two variables, apply the method of Lagrange Multipliers to determine the optimum of functions with constraints, and evaluate double and triple integrals.

Example 1: Find all relative extrema and saddle points of the function *f* defined by

$$f(x, y) = xy^2 - x^2y + 2xy + 5.$$

To simplify the format of subsequent commands, we will define *f* as an expression instead of as a function:

```
f := x * y^2 - x^2 * y + 2*x*y + 5;
```

The relative extrema and the saddle points can occur only at the critical points, and since *f* is a polynomial, these will only be the points at which $\frac{\partial f}{\partial x}$ and $\frac{\partial f}{\partial y}$ are both zero:

```
solve({diff(f, x) = 0, diff(f, y) = 0}, {x , y});
```

Maple may or may not be able to solve this system in general. In this case it can, producing the four solutions

(0, 0), (0, –2), (2, 0), and $\left(\frac{2}{3}, \frac{-2}{3}\right)$. Next, we define the discriminant $D = f_{xx} f_{yy} - (f_{xy})^2$

```
D := diff(f, x, x) * diff(f, y, y) - diff(f, x, y)^2;
```

And now we substitute the coordinates of each of the critical points into D:

```
subs( x = 0 , y = 0, D);
subs( x = 0 , y = -2, D);
subs( x = 2 , y = 0 , D);
subs( x = 2/3, y = -2/3, D);
```

Since D is negative, *f* has a saddle point at (0, 0), another saddle point at (0, –2), and another saddle point at (2, 0).

Since D is positive in this case, we must check the sign of f_{xx} at this point

```
subs(x = 2/3, y = -2/3, diff(f,x,x));
```

Since this is positive, we conclude that f has a relative minimum at $\left(\frac{2}{3}, \frac{-2}{3}\right)$.

Example 2: Use the method of Lagrange Multipliers to find the point on the line

$$2x - 3y + z = 1$$
$$x - 2y + 3z = 6$$

that is closest to the point (1, 4, 2).

Put differently, we want to find the minimum of

$$f(x, y, z) = (x - 1)^2 + (y - 4)^2 + (z - 2)^2$$

subject to the constraints

$$g_1(x, y, z) = 0 \text{ and } g_2(x, y, z) = 0, \text{ where}$$
$$g_1(x, y, z) = 2x - 3y + z - 1, \text{ and}$$
$$g_2(x, y, z) = x - 2y + 3z - 6.$$

We proceed by defining f, g_1, and g_2 as expressions:

```
f := (x - 1)^2 + (y - 4)^2 + (z - 2)^2;
g1 := 2*x - 3*y + z - 1;
g2 := x - 2*y + 3*z - 6;
```

Now define the expression T by

```
T := f - L1*g1 - L2*g2;
```

Here L_1 and L_2 are the Lagrange multipliers. If the maximum and minimum of f subject to the given constraints exist, they must satisfy the following conditions

$T_x = 0$, $T_y = 0$, $T_z = 0$, $g_1 = 0$, and $g_2 = 0$.

So we ask Maple to solve this system:

```
solve({diff(T, x) = 0, diff(T, y) = 0, diff(T, z) = 0, g1 = 0, g2
= 0}, {x, y, z, L1, L2} );
```

Maple gives us the unique answer

$$\left(\frac{172}{75}, \frac{31}{15}, \frac{196}{75}\right).$$

It is clear that this must correspond to the minimum of f, since the maximum cannot possibly exist (one could choose further and further points on the line).

Example 3: Use Maple to evaluate the double integral

$$\iint_R \frac{x^2 y}{x^4 + y^2}\, dA,$$

where R is the region bounded from above by the parabola $y = x^2$, from below by the x-axis, and from the sides by the lines $x = 0$ and $x = 1$.

This is clearly equivalent to the iterated integrals

$$\int_0^1 \int_0^{x^2} \frac{x^2 y}{x^4 + y^2}\, dy\, dx.$$

The corresponding Maple command is

```
int( int(x^2*y/(x^4 + y^2), y = 0 .. x^2), x = 0..1);
```

Maple gives the answer $\frac{1}{6}$ ln (2) in a few seconds.

Example 4: Use triple integrals to find the volume of the solid bounded by the paraboloid $z = 2x^2 + y^2$ and the cylinder $z = 2 - y^2$.

The two surfaces meet at a curve whose projection onto the xy-plane has equation $x^2 + y^2 = 1$. Thus the desired volume is just the triple integral

$$\int_{-1}^{1}\int_{-\sqrt{1-x^2}}^{\sqrt{1-x^2}}\int_{2x^2+y^2}^{2-y^2} dz\, dy\, dx.$$

We now ask Maple to evaluate this triple integral:

```
int (int (int (1 , z = 2*x^2 + y^2 .. 2 - y^2) ,
y = -sqrt(1 - x^2)..sqrt(1-x^2) ), x = -1..1);
```

Maple gives the exact answer of π.

Exercises with Maple

1*. Find all relative extrema and saddle points of the function

$$f(x, y) = 3x^2y + y^3 - 3x^2 - 3y^2 + 2.$$

....

2. Use the method of Lagrange Multipliers to find those points on the sphere $x^2 + y^2 + z^2 = 4$ that are nearest to and farthest from the point (2, 1, 2).

....

3. Express the area bounded by the ellipse $\dfrac{x^2}{a^2} + \dfrac{y^2}{b^2} = 1$ as a double integral and have Maple evaluate it.

....

....

A Maple Functions Used in This Book

1. Standard Functions

Function Name	**Example of Usage**
abs	x := abs(3 - sqrt(10));
add	add(u,v); or add(u, v, a, b); part of **linalg** package
angle	t := angle(u, v) * 180/Pi; part of **linalg** package
array	v := array([1, 2, 3]); or A := array([[1, 2, 3] , [4, 5, 6]]);
changevar	changevar(x^2 = u, Int(x*sqrt(9 - x^4), x), u); part of **student** package
completesquare	(x^2 + 3*x + 17, x); part of **student** package (in earlier releases, before Maple V, this is **cmpltsq**)
crossprod	w := crossprod(u , v); part of **linalg** package
convert	convert (taylor(f(x) , x = 0, n), polynom); or convert(x^2/(x^6 - 1), parfrac, x); **convert** has many other uses.
cylinderplot	cylinderplot([r , 3*pi/4 , z], r = 0 .. 5, z = -4 .. 4); Maple V only
denom	d := denom((x+1)/(x-1));
diff	diff(x^2*sin(x) , x); or diff(x^3 * tan(x*y) , x$4 , y$3);
display3d	display3d({G1 , G2}); Maple V only
dotprod	k := dotprod(u , v); part of **linalg** package
dsolve	dsolve({diff (y(t), t) = y(t) - 5 , y(0) = 2} , y(t));
evalc	evalc((1 + I)^15);
evalf	evalf(cos(2)); number of significant digits is 10, by default.
expand	expand((2*x + 3*y)^8);
factor	factor(x^ - 3*x + 2);
fsolve	fsolve(x^3 - 4*x^2 - 12*x - 7 = 0);
gcd	gcd((x^2 - x - 2 , x^2 - 1)); or gcd(120 , 84);
ifactor	ifactor(2^35 - 1);
int	int(x*sin(x) , x);
Int	The inert form of int. Maple does not evaluate the integral.
intparts	intparts(Int (x*sin(x) , x), sin(x)); part of **student** package
limit	limit(sin(x)/x , x = 0); or limit(x*y/(x^2 + y^2) , {x = 0 , y = 0});
linsolve	linsolve(A , b); part of **linalg** package
normal	normal ((x^2 - 1)/(x - 1));

numer	d := numer((x+1)/(x-1));
plot	plot(x*sin(x) - x - 2 , x= -3 .. 3, y =-5 .. 5); or
	plot({sin(x) , cos(x)}); (simultaneous plotting), or
	plot([2*cos(t) , 2*sin(t), t = 0 .. 2*Pi], -5 .. 5, -3 .. 3); (parametric), or
	plot([3*sin(t/2), t , t = 0 .. 4*pi], -5 .. 5, -3 .. 3, coords = polar); (polar)
plot3d	plot3d(sin(sqrt(x^2 + y^2)) , x = -6.. 6, y= -6 .. 6); Maple V only
radsimp	radsimp((x^2 - 4*x + 4)^(1/2));
read	read(`Newton.m`);
simplify	simplify(9^(1/2) - arccos(1/2));
solve	solve (x^3 - 3*x^2 + 7*x - 5, x); or
	solve ({3*x + 19*y = 14, 15*x - 11*y = 169}, {x , y});
sphereplot	sphereplot([r , t , Pi/6] , r = 0 .. 4, t = 0 .. 2*Pi); Maple V only
sqrt	d := sqrt(x^2 + y^2 + z^2);
subs	subs(x = 2/3, y = -7/8, 2*x^2 - y^2 + 12*x - 5*y +126);
sum	sum(k^4 - 5*k^2 + 120*k -5, k = 1 .. 12345); or
	sum(1/k^2 , k = 1 .. infinity); or
	sum(k^3 , k = 1 .. n);
taylor	taylor(arcsin(x), x = 0, 43);
with	with(student); or with(linalg);

The standard mathematical functions include **sin()**, **cos()**, **tan()**, **cot()**, **sec()**, **csc()**, **sinh()**, **cosh()**, **tanh()**, **coth()**, **sech()**, **csch()**, **arcsin()**, **arccos()**, **arctan()**, **arccot()**, **arcsec()**, **arccsc()**, **exp()**, **ln()**, and many more. For further information use Maple's Help menu.

2. Nonstandard Functions

The following functions can be invoked only after they have been read using a read statement. It is assumed that the m-files containing the codes for these functions have been included in the Maple folder. All these functions assume that f has already been defined as a function

Function Name	Example of Usage
Der	Der(f, 1);
LeftSum	LeftSum(f, 1 , 2, 40);
LimTable	LimTable(f , 1); or LimTable(f , infinity);
limtab	Same as LimTable; more appropriate with very small numbers.
Newton	Newton(f , 1 , 2, 10);
RightSum	RightSum(f , 1, 2, 40);
Simpson	Simpson(f , 1 , 2 , 40);
SumTable	SumTable(f , 1 , 2);
Trapezoid	Trapezoid(f , 1 , 2 , 40);

The functions **LeftSum, RightSum, and SumTable** are all available with the command **read(`Sums.m`)**, and **limtab** is available when **LimTable** has been read.